D0999724

UNIT OPERATIONS MODELS FOR SOLID WASTE PROCESSING

UNIT OPERATIONS MODELS FOR SOLID WASTE PROCESSING

by

G.M. Savage, J.C. Glaub, L.F. Diaz
Cal Recovery Systems, Inc.
Richmond, California

Library of Congress Catalog Card Number 86-5154
ISBN: 0-8155-1086-1
ISSN: 0090-516X
Printed in the United States

Published in the United States of America by
Noyes Publications
Mill Road, Park Ridge, New Jersey 07656

10 9 8 7 6 5 4 3 2 1

Library of Congress Cataloging-in-Publication Data

Savage, George M., 1948–
Unit operations models for solid waste processing.

(Pollution technology review ; no. 133)
Bibliography: p.
Includes index.
1. Refuse and refuse disposal--Mathematical models--Handbooks, manuals, etc. 2. Factory and trade waste--Mathematical models--Handbooks, manuals, etc. I. Glaub, J.C. II. Diaz, Luis F. III. Title. IV. Series.
TD793.S27 1986 628.4'45'0724 86-5154
ISBN 0-8155-1086-1

Foreword

This book presents models for the key unit operations used in solid waste front-end processing systems. Models are provided for the mass balance, energy requirements, and economics of the following operations:

- Size Reduction
- Air Classification
- Trommel Screening
- Ferrous Separation
- Non-Ferrous Separation
- Glass Separation
- Disc Screening
- Shear Shredding
- Densification
- Conveying

In keeping with the objective of ultimately assembling the various unit operation models into a resource recovery system model, the formats of the inputs and outputs of each model have been structured in a form that is compatible. The models have been developed from first principles whenever possible, and invoke empiricism in those cases where a governing theory has yet to be developed. Field test data and manufacturers' information, where available, supplemented the analytical development of the models.

Within the resource recovery industry, there is a need for unit operation models which, when integrated into a system, would allow an evaluation of the system design. System design is taken to include the determination of the yields and quality of the different flow streams as well as the associated system costs as a function of various operating conditions. Model development in the past has been of such a form as to preclude an easy, straightforward integration of different unit operation models into an overall system model (i.e., aggregation of separate unit operation models). Consequently no generalized integrated system model exists for front-end processing. Those that do exist are limited to single-system configurations and do not integrate the mass balance, energy requirements, and cost algorithms into a unified system model.

The book contains four parts—a review of the unit operation modeling literature, plus sections on mass balance and energy requirement models, economic models, and unit operation models. In addition to the description of the mathematical equations that describe each unit operation, the geneology of each unit operation model is discussed briefly along with the assumptions and limitations placed upon the models. In the case of several of the unit operations, the lack of an accepted governing theory, field test data, or both, limited the development of the models. In those cases, empiricism and best engineering judgement were used to formulate a satisfactory model. This book should be of interest to engineers and scientists developing solid waste processing projects and programs.

The information in the book is from *Models of Unit Operations Used for Solid Waste Processing,* prepared by G.M. Savage, J.C. Glaub, and L.F. Diaz of Cal Recovery Systems for the U.S. Department of Energy, Argonne National Laboratory, September 1984.

The table of contents is organized in such a way as to serve as a subject index and provides easy access to the information contained in the book.

Advanced composition and production methods developed by Noyes Publications are employed to bring this durably bound book to you in a minimum of time. Special techniques are used to close the gap between "manuscript" and "completed book." In order to keep the price of the book to a reasonable level, it has been partially reproduced by photo-offset directly from the original report and the cost saving passed on to the reader. Due to this method of publishing, certain portions of the book may be less legible than desired.

NOTICE

Contents and Subject Index

PART II
TASK 3 REPORT: MASS BALANCE AND
ENERGY REQUIREMENT MODELS

PART III
TASK 4 REPORT: ECONOMIC MODELS

PART IV
TASK 5 REPORT: UNIT OPERATION MODELS: MASS BALANCE, ENERGY REQUIREMENTS, AND ECONOMICS

Introduction

Within the resource recovery industry, there is a need for unit operation models which, when integrated into a system model, would allow an evaluation of the system design. System design is taken to include the determination of the yields and quality of the different flow streams as well as the associated system costs as a function of various operating conditions. Model development in the past has been of such a form as to preclude an easy, straightforward integration of different unit operation models into an overall system model (i.e., aggregation of separate unit operation models). Consequently, no generalized integrated system model exists for front-end processing. System models that do exist are limited to single-system configurations and do not integrate the mass balance, energy requirements, and cost algorithms into a unified system model.

The main body of the report presents mass balance, energy requirement, and economic models for the following unit operations:

- Size Reduction
- Air Classification
- Trommel Screening
- Ferrous Separation
- Non-Ferrous Separation
- Glass Separation
- Disc Screening
- Shear Shredding
- Densification
- Conveying

In addition to the description of the mathematical equations that describe each unit operation, the geneology of each unit operation model is discussed briefly along with the assumptions and limitations placed upon the models. In the case of several of the unit operations, the lack of an accepted governing theory, field test data, or both limited the development of the models. In those cases, empiricism and our best engineering judgment were used to formulate a satisfactory model. Among the unit operations that were modeled and for which theory and information are lacking are numbered the shear shredder, disc screen, non-ferrous separator, and glass separator.

Cal Recovery Systems, Inc., under contract to Argonne National Laboratory, was commissioned to develop state-of-the-art models for the unit operations that typically comprise front-end processing systems. The models, which have been developed and are presented here, represent the mass balance, energy requirements, and economics of the unit operations and are derived where possible from basic principles. In addition, for the purpose of allowing the incorporation of the unit operation models into a system model, the unit operation models have been structured so that the form of the inputs and the outputs are compatible. The means for accomplishing the latter goal consists of structuring the inputs and outputs in a matrix format consisting of the types of components (i.e., ferrous, glass, paper, etc.) and their respective size classes. The above format structure greatly increases the complexity and sophistication of the models and provides the basic structure for allowing comprehensive system analysis in the true engineering sense, i.e., system dynamics and performance can be assessed as a function of operating conditions and feedstock composition.

Also included in this report is a literature review for the purpose of compiling and discussing in one document the available information pertaining to the modeling of front-end unit operations.

This report is an assemblage of four task reports. The Task 2 Report presents the review of the literature. The Task 3 Report describes the mass balance and energy requirement models for the most commonly employed and the most studied unit operations in front-end processing, namely size reduction, air classification, trommel screening, ferrous separation, densification, and conveying. The Task 4 Report presents the economic models for the unit operations described in the Task 3 Report. In the Task 5 Report are described models for four of the less commonly employed and less studied unit operations, namely shear shredding, disc screening, glass separation, and non-ferrous separation. The Task 5 Report presents models for the mass balance, energy requirements, and economics of the above unit operations.

Part I

Task 2 Report: Review of the Literature: Modeling Unit Operations for RDF Processing

1. Introduction

The literature review was undertaken to secure information and data that would help to form a basis for the development of models for the various unit operations used in RDF processing. The acquired information will serve to complement the substantial in-house information and field test data that CRS already had on hand at the initiation of the modeling project.

Moreover, the review represents a compilation of literature whose contents are applicable to the various aspects that bear upon the development of models for unit operations. The aspects include the description of equipment and flow diagrams of specific processing sequences that compose a given generic unit operation, the development of the basic engineering principles and governing equations for describing specific unit operations, and the acquisition of performance data and of cost information for the purpose of verifying the validity of the models.

The literature review was specifically concerned with the above aspects and consequently is not intended to be a comprehensive literature review in the usual sense. The review is structured to allow readers to locate, in an expeditious manner, published information on subjects germane to the modeling of unit operations.

For particular unit operations (e.g., non-ferrous and glass separation), there is a substantial lack of technical engineering detail. In those cases, the literature review describes the available information, even though it may be of a general nature.

The review of the literature on RDF processing operations is organized under the following topics:

- Size Reduction
- Air Classification
- Screening
- Densification
- Ferrous Separation
- Non-Ferrous Separation
- Glass Separation
- Conveying

2. Size Reduction

The early research in refuse comminution for the most part was along the lines of that followed in the mineral processing industry. The tendency of this type of research was to treat MSW as a homogeneous and brittle material as a first-order approximation. Among those active in research in the mineral processing industry were Bond [1], Austin and Kimpel [2], Charles [3], Schumann [4], Austin and Luckie [5], and Austin, Luckie, and Klimpel [6]. The fact that only about 25 percent of MSW is composed of brittle material was not taken into account at first because no theories pertaining to the size reduction of non-brittle materials were to be found in the mineral waste literature. The reason that the "brittle material" theories cannot be successfully extended to MSW most likely is the fact that MSW is a heterogeneous mixture of brittle and non-brittle materials.

Literature dealing specifically with the modeling and characterization of refuse size reduction dates from the research conducted by Trezek, Savage, Shiflett, and Obeng at the University of California (Berkeley) in the early 1970's. The researchers directed their investigations towards the identification of the key parameters that characterize refuse size reduction and their methods of measurement [7,8,9]. They developed mathematical relationships pertaining to degree of size reduction, energy requirements, hammer wear, and moisture content in the size reduction of raw MSW. Through the use of process modeling techniques developed in the field of mineral processing and by the incorporation of the concepts of selection and breakage functions, Obeng [10] sought to predict refuse particle breakage by way of four different matrix models. Variations in feed rate, moisture content, and number of stages of size reduction were simulated by Obeng on the basis of data gathered in a 10 Mg/hr swing hammermill shredder. The matrix method developed by Obeng was subsequently extended by Shiflett [11] through the use of non-linear optimization techniques and the application of the concept of residence time to determine values for the selection and breakage functions for raw MSW. Shiflett relied upon the 10 Mg/hr hammermill upon which Obeng based his work.

Neither Obeng's nor Shiflett's research made an allowance for correlation of product size distribution with the energy required for size reduction or with the internal configuration of the size reduction equipment. Shiflett and Trezek [12] reported a first-order attempt to reconcile production size distributions with variations in mean residence times by introducing the concept of material "holdup" in the size reduction device and incorporating the flowrate of material through the machine.

Results obtained by Savage and Trezek [13] in their investigation of the influence of particle size and composition of the refuse feedstock indicated that the specific energy increased markedly with an increase in the particle size of the feedstock and the percentage of cellulosic material. The influence of the particle size of the feedstock and of the feedstock composition on specific energy requirements were also investigated by Alter and Stratton [14] for certain materials (e.g., tin cans, aluminum cans). Other than the reports by Savage and Trezek and by Alter and Stratton, the literature is devoid of information on the comminution of refuse components and on the effect of feedstock size. Excepting the paper by Trezek, Savage, and Howard in Compost Science [15], no information on the basic mechanical properties of refuse components is readily available.

In contrast with the dearth of theoretical descriptions of the refuse comminution process, several papers are available that deal with studies aimed at documenting the performance of commercial refuse size reduction equipment. For example, Alter and Stratton [14] reported on the performance of 11 different refuse shredders. Savage and Shiflett [16] gathered and reported field test data on the product size, energy requirements, and hammer wear for nine size reduction machines. In a follow-up study, Savage, et al [17] examined the influence of machine parameters (number of hammers, etc.) and number of size reduction stages on product size, hammer wear, and energy requirements in the size reduction of refuse. Vesilind and Rimer [18] have reported on the performance of a vertical hammermill in terms of product size characterization, hammer wear, and energy requirements. Ruf [19] has analyzed the size distributions of refuse milled with a hammermill and with a rasp mill. Ham and Reinhart [20] have determined hammer wear and product size distributions for a vertical and for a horizontal hammermill.

Product size distributions, hammer wear, energy requirements, moisture content, and the composition and particle size of the refuse feedstock have been identified during the literature search as key parameters that characterize refuse size reduction. An overview and a discussion of the five parameters along with engineering design information and cost data on refuse size reduction are given by Savage, et al in an engineering manual authored by them [21]. An overview of the fundamentals of refuse size reduction is also presented in two books, one authored by Diaz, et al [22] and the other by Vesilind and Rimer [23].

2.1 REFERENCES

1. Bond, F.C., "The Third Theory of Comminution," Transactions AIME, 193:484-494 (1952).
2. Austin, L.G., and R.R. Klimpel, "Theory of Grinding Operations," I-EC Process Design and Development, 56:19-29 (1964).
3. Charles, R.J., "Energy-Size Reduction Relationships in Comminution," Transactions AIME, 208:80-88 (1957).
4. Schumann, R., "Energy Input and Size Distribution in Comminution," Transactions AIME, 73:22-25 (1960).
5. Austin, L.G., and P.T. Luckie, "Grinding Equations and the Bond Work Index," Transactions AIME, 252 (September 1972).
6. Austin, L.G., P.T. Luckie, and R.R. Klimpel, "Solutions of the Batch Grinding Equation Leading to Rosin-Rammler Distributions," Transactions AIME, 252 (March 1982).
7. Trezek, G.J., Significance of Size Reduction in Solid Waste Management, EPA-600/2-77-131 (July 1977).
8. Trezek, G.J., and G.M. Savage, Significance of Size Reduction in Solid Waste Management, Vol. 2, EPA-600/2-80-115 (August 1980).
9. Trezek, G.T., D.M. Obeng, and G.M. Savage, Size Reduction in Solid Waste Processing, Second Year Progress Report - 1972-1973, EPA Grant No. R801218.
10. Obeng, D.M., "Comminution of a Heterogeneous Mixture of Brittle and Non-Brittle Materials," Ph.D. dissertation, University of California, Berkeley (1973).
11. Shiflett, G.R., "A Model for the Swing-Hammermill Size Reduction of Residential Refuse," D. Eng. dissertation, University of California, Berkeley (1978).

12. Shiflett, G.R., and G.J. Trezek, "The Use of Residence Time and Nonlinear Optimization in Predicting Comminution Parameters in the Swing Hammermilling of Refuse," I-EC Process Design and Development, 18(3):437-440 (1979).

13. Savage, G.M., and G.J. Trezek, Significance of Size Reduction in Solid Waste Management, Volume II, EPA-600/2-80-115 (August 1980).

14. Stratton, F.E., and H. Alter, "Application of Bond Theory to Solid Waste Shredding," Journal Environmental Engineering Division, ASCE, 104 (EEI) (1978).

15. Trezek. G.J., D. Howard, and G.M. Savage, "Mechanical Properties of Some Refuse Components," Compost Science, 15(4) (November-December 1972).

16. Savage, G.M., and G.R. Shiflett, Processing Equipment for Resource Recovery Systems, Vol. III, Field Test Evaluation of Shredders, EPA-600/2-80-007c (July 1980).

17. Savage, G.M., J.K. Tuck, P.A. Gandy, and G.J. Trezek, Significance of Size Reduction in Solid Waste Management, Volume 3 - Effects of Machine Parameters and Shredder Performance, EPA Contract No. 68-03-2866 (August 1982).

18. Vesilind, P.A., A.E. Rimer, and W.A. Worrell, "Performance Characteristics of a Vertical Hammermill Shredder," Proceedings of the 1980 ASME National Waste Processing Conference (May 1980).

19. Ruf, J.A., "Particle Size Spectrum and Compressibility of Raw and Shredded Municipal Refuse," University of Florida (1974).

20. Ham, R.K., and J.J. Reinhart, Final Report on a Demonstration Project at Madison, Wisconsin, to Investigate Milling of Solid Wastes Between 1966-1972, Vol. I, EPA Grant No. 3-G06-EC-00000-0051.

21. Savage, G.M., D.J. Lafrenz, D.B. Jones, and J.C. Glaub, Engineering Design Manual for Solid Waste Size Reduction Equipment, EPA Contract No. 68-03-2972 (September 1982).

22. Diaz, L.F., G.M. Savage, and C.G. Golueke, Resource Recovery from Municipal Solid Wastes, Volume I, CRC Press (1982).

23. Vesilind, P.A. and A.E. Rimer, Unit Operations in Resource Recovery Engineering, Prentice-Hall, Inc. (1981).

3. Air Classification

Most of the available literature on air classification consists of reports on experimental work. For example, Senden [1] has investigated and reported the effects of chamber geometry and air velocity on the separation process. Murray and Liddell [2] examined the effects of moisture content and particle size of the air classifier-feed and of the feed rate on performance as expressed in terms of percent lights recovered. In a paper presented at the Eighth Biennial National Waste Processing Conference, Murray [3] gives the efficiency of recovery of the light fraction for two tests carried out at the same input moisture content, particle size, and flowrate. In their investigation, Musil and Scholz [4] used two 1/4-scale model air classifiers (a zigzag and a rotary drum) to compare experimental fluidizing velocities of different refuse components with those predicted by theory. The important operating variables listed by the authors were material feed rate and air velocity. With respect to the rotary drum air classifier tested, Musil and Scholz identified angle of inclination and rotational speed as also being important parameters. Worrell and Vesilind [5] determined the variation of air classifier efficiency with air speed for three different throat designs. For their study, Worrell and Vesilind defined efficiency as being the product of the fraction of light material in the overflow and the portion of heavies in the underflow. This definition was also applied in a study by Vesilind and Henrikson [6] on the effect of feed rate on efficiency.

In a comprehensive study on the performance of seven full-scale air classifiers, Hopkins, et al [7] investigated the ability of each unit to concentrate light and heavy material in the respective streams at different air-to-solids (infeed) ratios, and the energy consumed in doing so. Other operating parameters investigated by the researchers were input feed rate and average column velocity. Capital costs of the tested systems are given.

The modeling of the air classification process is the subject of two publications. Senden and Tels [8] present a mathematical model that can be used for deriving a relationship between separation efficiency and the mean particle residence time in the air classifier. The model was later modified by Henrikson [9]. In Henrikson's modification, the properties of the air classified light fraction (moisture content, ash content, and heating value) are expressed in terms of air velocity. Yet another model, i.e., a light-fraction model, has been proposed by Fan [10].

In their book, Diaz, et al [11] provide an in-depth review of air classification that covers design and operational factors and air classifier costs. Among the important operational variables listed by the authors are air-to-solids ratio, column loading (defined as material feed rate divided by the column area), and air velocity. In the book, cost information based upon data gathered by Chrisman [12], Even, et al [13], and Fiscus, et al [14] is given for seven air classifier systems. In their publication, Vesilind and Rimer [15] present graphs that show air classifier efficiency, percent of material reporting to the light fraction, and value of recovered materials as functions of air velocity. Saheli [16] points out that the important air classifier design variables are particle size and shape, air-to-solids ratio, particle terminal velocity, and the ratio of solid density to fluid density.

The key operational parameters commonly identified in the literature are: solid particle size and shape, feed rate, air-to-solids ratio, air velocity, and column geometry.

3.1 REFERENCES

1. Senden, M.M.G., "Performance of Zigzag Air Classifiers at Low Particle Concentrations: A Study of the Effect of Stage Geometry Variations," ASME, Proceedings of the Ninth Biennial National Waste Processing Conference (1980).

2. Murray, D.L., and C.R. Liddell, "The Dynamics, Operation and Evaluation of an Air Classifier," Waste Age, 8:18 (March 1977).

3. Murray D.L., "Air Classifier Performance and Operating Principles," ASME, Proceedings of the Eighth Biennial National Waste Processing Conference (1978).

4. Musil, J.E., and P.D. Scholz, "Reduced Size Model Study of Two Air Classifiers," Journal of the Environmental Engineering Division, p. 659 (August 1978).

5. Worrell, W.A., and P.A. Vesilind, "Testing and Evaluation of Air Classifier Performance," Resource Recovery and Conservation, 4:247 (1979).

6. Vesilind, P.A. and R.A. Henrickson, "Effect of Feedrate on Air Classifier Performance, Resources and Conservation, 6:211 (1981).

7. Hopkins, V., et al, Comparative Study of Air Classifiers, Midwest Research Institute, Kansas City, MO., EPA Contract No. 68-03-2730 (1980).

8. Senden, M.M.G., and M. Tels, "Mathematical Model of Vertical Air Classifiers," Resource Recovery and Conservation, 3:129 (May 1978).

9. Henrikson, R., Analytical Evaluation of Air Classification, Duke Environmental Publication, Duke University (1979).

10. Fan, D., "On the Classified Light Fraction of Shredded Municipal Solid Waste," Resource Recovery and Conservation, 1:141 (1975).

11. Diaz, L.F., et al, Resource Recovery from Municipal Solid Wastes, Volume I, CRC Press, Inc., Boca Raton, Florida (1982).

12. Chrisman, R.L., Air Classification in Resource Recovery, National Center for Resource Recovery, Inc., RM78-1 (October 1978).

13. Even, J.C., et al, Evaluation of the Ames Solid Waste Recovery System I," U.S. EPA, Cincinnati, Ohio (1977).

14. Fiscus, D.E., et al, St. Louis Demonstration Final Report, EPA-600/12-77-155a, U.S. EPA, Cincinnati, Ohio (September 1977).

15. Vesilind, P.A., and A.E. Rimer, Unit Operations in Resource Recovery Engineering, Prentice-Hall, Inc., Englewood Cliffs, NJ (1981).

16. Saheli, F.P., "The Technology of Solid Waste Air Classification for Resource Recovery," ASME 76-ENAS-50 (1976).

4. Screening

4.1 BACKGROUND

Three types of screens have been employed in resource recovery applications, namely, flatbed screens, trommel screens, and disc screens. Accordingly, the literature review on screening is structured in terms of these three basic screen types. With regard to waste processing, most publications on screening are on trommel screens.

Analytical work on screening applications in waste processing has been reported in only a few publications. Reports on experimental and field test work on screens used in waste processing can be found in a somewhat larger number of publications, although the number is nevertheless relatively small. However, relevant publications in the general literature on other screening applications and processes can be of value in modeling resource recovery screening processes. Accordingly, the literature review presented herein encompasses some key publications that concern other screening applications and processes that can be related to screening applications in waste processing.

A comprehensive literature review of screening, including individual summaries of the various publications, has been made by Glaub, *et al* [1].

4.2 FLATBED SCREENING

The only publication regarding the use of flatbed screens in waste processing is a paper by Savage and Trezek [2]. In their work, the efficiencies of screening shredded, air-classified light fraction with a flatbed screen were experimentally determined and compared to those obtained with a trommel screen. A range of mass flowrates were tested, and several modifications to the flatbed screen were examined. It was found that the trommel screen substantially outperformed the flatbed screen. Modifications to improve the performance of the flatbed screen did not meet with success. In addition to the poorer performance, power consumption was greater with the flatbed screen.

A number of analytical models regarding flatbed screening processes are presented in the literature. Although none of these publications are directly concerned with screening municipal solid waste, most of the relations developed are broadly applicable. That is not to say that the theoretical models conform well with experimental results for all materials. Moreover, many of the models only apply to certain shapes of materials (e.g., spherical or granular) or to certain conditions in the screening process (e.g., near the end of the screening process or the presence of a well-mixed bed). Particle dynamics models for flatbed screening processes are presented in Refs. 3 to 5. Screening rate models for flatbed screening processes are presented in Refs. 5 to 12.

4.3 TROMMEL SCREENING

The technical literature contains few publications on the utilization of trommel screens in waste processing. The majority of such publications are experimental in nature, and only a few deal with the analytical modeling of the trommel screening process. Moreover, very few publications on other applications of trommel screening have appeared in the general literature.

A theoretical model for the screening rate in a trommel screen was presented by Bodziony in 1960 [13]. According to Bodziony's model, the screening rate for a given size fraction was equal to a screening rate coefficient for that size fraction times the volume fraction of particles in the fraction. No particle dynamics model was presented. Vorstman and Tels presented a similar model, but altered the term for volume fraction to surface concentration [14]. Their work was very much concerned with the occurrence of particle segregation (i.e., demixing) in the trommel screening process. Unlike the typical operation of trommel screens in the U.S. in which the material is lifted and dropped, Vorstman and Tels were concerned with the operation of trommel screens in kiln action. A particle dynamics model was not presented by the two researchers.

Models for screening rate and particle dynamics were described in a paper presented by Alter, et al at the ASME 9th National Waste Processing Conference [15]. Their screening rate model was based on a simple probability of passage model that had been presented by many other researchers. Their particle dynamics model was based on a model developed by Davis for ball mills [16].

More recently, in a report to DOE, Glaub, et al [1,17] have developed and described models for screening rate and particle dynamics. On the basis of a probability of passage model proposed by Gaudin [7] that accounts for reflection of particles through apertures in a wire-mesh screen, Glaub, et al, developed a probability of passage model in which the reflection on a punched-plate screen is taken into account. (It should be noted that all existing probability of passage models (including simple and reflection models) have the shortcoming of being based upon a spherical particle geometry.) Glaub, et al, also proposed a particle dynamics model that accounts for slippage of material on the screen surface, and then proceeded to confirm the existence of slippage through the use of photographic experiments. In their report to the DOE [1], Glaub, et al describe the results of a variety of trommel screening experiments, probability of passage experiments, and measurements of material properties required as inputs to trommel screening models.

Transfer function models for screening processes were presented by Trezek, et al in a report to the Electric Power Research Institute [18]. Their models include transfer functions for various waste stream components and are given for pre-trommeling and post-trommeling processes.

Experimental data on trommel screening can be found in several publications. Of value to modeling the screening process are those data that regard component and size mass balances. In Ref. 1, Glaub, et al present component and size mass balances for trommel screening raw MSW and shredded, air-classified light fraction. Hennon, et al, give detailed component and size mass balances for 60 field tests of a trommel screen [19]. They conducted the tests, on a quarterly basis, over a one-year period at the Baltimore County resource recovery facility. The trommel was utilized to screen shredded, air-classified light fraction. Trezek and Savage report component and size mass balances that were based upon test data obtained at the University of California's Richmond Field Station [20].

Bernheisel, et al, report on the component mass splits obtained in two test runs with a pre-trommel at the resource recovery facility in New

Orleans [21]. Campbell presented component and size mass balances for five additional test runs with the New Orleans pre-trommel [22].

In a report on work performed for the DOE, Barton presents the results of field tests on two commercial trommel screens in England [23]. One trommel was used to screen raw refuse, and the other trommel was used to screen shredded refuse. Component and size mass balances are reported for the tests on one of the screens. Only the size distributions of the splits are reported for the second screen. The unusual aperture shapes and configurations of the screens render the reported data of little use to modeling efforts.

Limited mass balance data also are presented by Woodruff [24] and Woodruff and Bales [25]. Other publications in which are reported results obtained in trommel screening experiments are those by Savage and Trezek [2], Douglas and Birch [26], and Warren [27].

4.4 DISC SCREENING

There is a paucity of information in the technical literature on disc screening. Fiscus, *et al* [28] have reported the results obtained in a field test conducted at the Ames resource recovery plant. Unfortunately, they did not collect enough samples to develop a mass balance for the screen or even to determine screening efficiency. Instead, fuel characteristics and characteristics of the inputs to downstream unit processes are compared before and after the disc screen was installed. A qualitative discussion of how a disc screen functions was presented by Hamilton and Kelyman in *Solid Wastes Management* [29].

4.5 REFERENCES

1. Glaub, J.C., D.B. Jones, J.U. Tleimat, and G.M. Savage, *Trommel Screen Research and Development for Applications in Resource Recovery*, Final Report under U.S. DOE Contract No. DE-AC03-79CS20490 (October 1981).

2. Savage, G. and G.J. Trezek, "Screening Shredded Municipal Solid," *Compost Science* (January/February 1976).

3. Davis, R.F., "The Dynamics of Screening," *Transactions, Institution of Chemical Engineers*, *18*:76-79 (1940).

4. Wolff, E.R., "Screening Principles and Applications," Industrial and Engineering Chemistry, 46:1778-1784 (1954).

5. Jansen, M.L., and J.R. Glastonbury, "The Size Separation of Particles by Screening," Powder Technology, 1:334-343 (1967/68).

6. Wiard, E.S., "The Grading Industries - II," Metallurgical and Chemical Engineering, XIV(4):191-197 (1916).

7. Gaudin, A.M., Principles of Mineral Dressing, McGraw-Hill, New York, (1939).

8. Estridge, R., "Initial Retention of Elongated Particles on Idealized Screens," Industrial and Engineering Chemistry-Process Design and Development, 1:87-91 (1962).

9. Kaye, B.H., "Investigation Into the Possibilities of Developing a Rate Method of Sieve Analysis," Powder Metallurgy, 10:199-217 (1962).

10. Baldwin, P.L., "The Continuous Separation of Solid Particles by Flat Deck Screens," Transactions, Institution of Chemical Engineers, 41:255-263 (1963).

11. Hudson, R.B., M.L Jansen, and P.B. Linkson, "Batch Sieving of Deep Particulate Beds on a Vibratory Sieve," Powder Technology, 2:229-240, (1968/69).

12. King, E.H., "How to Determine Plant Screening Requirements," Chemical Engineering Progress, 73(5):74-79 (1977).

13. Bodziony, J., Bulletin de l'Acadamie Polonaise des Sciences, 8:99-106, (1960).

14. Vorstman, M.A.G. and M. Tels, "Segregation in Trommelsieving," Recycling Berlin '79, Volume II (1979).

15. Alter, H., J. Gavis, and M. Renard, "Design Models of Trommels for Resource Recovery Processing," Proceedings of the 9th National Waste Processing Conference, American Society of Mechanical Engineers, Washington, D.C. (May 1980).

16. Davis, E.W., "Fine Crushing in Ball-Mills," Transactions, American Institute of Mining and Metallurgical Engineers, 61:250-296 (1919).

17. Glaub, J.C., D.B. Jones, and G.M. Savage, "The Design and Use of Trommel Screens for Processing Municipal Solid Waste," Proceedings of the 10th National Waste Processing Conference, American Society of Mechanical Engineers, New York City (May 1982).

18. Trezek, G.J., L.F. Diaz, G.M. Savage, and R. White, Prediction of the Impact of Screening on Refuse-Derived Fuel Quality, Report No. FP-1249, Electric Power Research Institute, Palo Alto, California (1979).

19. Hennon, G.J., D.E. Fiscus, J.C. Glaub, and G.M. Savage, An Economic and Engineering Analysis of a Selected Full-Scale Trommel Screen Operation, Final Report, Department of Energy Contract No. DE-AC03-80CS24330 (1983).

20. Trezek, G.J., and G.M. Savage, "MSW Component Size Distributions Obtained from the Cal Resource Recovery System," Resource Recovery and Conservation, 2:67-77 (1976).

21. Bernheisel, J.F., P.M. Bagalman, and W.S. Parker, "Trommel Processing of Municipal Solid Waste Prior to Shredding," Proceedings of the Sixth Mineral Waste Utilization Symposium, W.S. Bureau of Mines/Illinois Institute of Technology, Chicago (May 1978).

22. Campbell, J., New Orleans Full Scale Trommel Screen Evaluation: Interim Test Report, Draft Report, U.S. Department of Energy Contract No. DE-AC03-80CS-24315 (October 1981).

23. Barton, J., Evaluation of Trommels for Waste to Energy Plants, Phase 1 Report of the Doncaster and Byker Test Series, Draft Report, U.S. Department of Energy Contract No. DE-AC03-80CS-24315 (December 1981).

24. Woodruff, K.L., "Preprocessing of Municipal Solid Waste for Resource Recovery with a Trommel," Transactions, Society of Mining Engineers, 260:201-204 (1976).

25. Woodruff, K.L., and E.P. Bales, "Preprocessing of Municipal Solid Waste for Resource Recovery with a Trommel-Update 1977," Proceedings of the 8th National Waste Processing Conference, American Society of Mechanical Engineers, Chicago (May 1978).

26. Douglas, E., and P.R. Birch, "Recovery of Potentially Re-Usable Materials from Domestic Refuse by Physical Sorting," Resource Recovery and Conservation, 1:319-344 (1976).

27. Warren, J.L., "The Use of a Rotating Screen as a Means of Grinding Crude Refuse for Pulverization and Compression," Resource Recovery and Conservation, 3:97-111 (1978).

28. Fiscus, D.E., A.W. Joensen, A.O. Chantland, and R.A. Olexsey, "Evaluation of the Performance of the Disc Screens Installed at the City of Ames, Iowa Resource Recovery Facility," Proceedings of the 1980 National Waste Processing Conference, American Society of Mechanical Engineers (1980).

29. Hamilton, F. and J. Kelyman, "Disc Screen," Solid Wastes Management, (May 1979).

5. Densification

The need to densify biomass was encountered by the agricultural community long before it was felt by the producers and users of RDF. Consequently, the early work on densification was done by agricultural engineers who were seeking means for facilitating the handling and storing of animal feedstuffs. The limited relevance of their findings to the densification of RDF is due to the fact that densified animal feed is significantly different from RDF in several important aspects, among which are the following:

1. Feedstuffs are mainly mixtures of grass, ground grain, vegetable oils, and molasses. Under pressure, these materials tend to flow more easily than do paper and plastic -- the main constituents of RDF. Furthermore, molasses and other nutrient additives act as natural binding agents for the densified feedstuff.
2. The density of a feedstuff pellet is equivalent to about 480 to 770 kg/m^3. At higher densities, the pellets are more difficult to masticate by the animals. RDF, on the other hand, must be densified to about 960 kg/m^3 in order to ensure sufficient structural integrity.

Despite the differences between the two classes of materials, some of the densification machinery used in agriculture (e.g., rotary disc and roller type pellet mills) has been successfully used for RDF densification with little or no modification. Consequently, research on densification conducted by the agricultural community is of interest. However, it should be pointed out that the research has tended to be aimed at producing results for only a few sets of operating conditions and feed mixtures, rather than at developing a unified theory of the densification process.

Bellinger and McColly [1] found that in densifying alfalfa hay into a closed cylinder, an increase in the moisture content from 4.6 to 21.1 percent resulted in a decrease in the compressive energy from 3.9 kWh/Mg to 2.8 kWh/Mg.

Pfost [2], working with turkey- and chicken-feed, found that the energy required to form pellets was from 3.7 to 5.3 kWh/Mg. His attempts to correlate the temperature of the die with energy consumption were of limited value, in that the range of temperatures attained was from 80°F to 125°F -- much less than that encountered under production-scale conditions with RDF. The addition of a calcium-based binding agent increased the durability of the pellets and decreased energy requirements.

Dobie [3] observed that the pelletization of grasses and straw resulted in a reduction of the roughage factor -- a development that is desirable in feed for ruminants. The reduction in roughage is a result of the substantial size reduction that occurs in pelletizing. He concluded that high quality cubes could be formed through compression and with the aid of various binding agents such as ammonium lignin sulfonate. (He used a closed end cylinder in his study.)

The densification of processed refuse was modeled by Ruf [4]. He found that in the range of about 0 to 1400 kPa, the density of typical refuse in Gainesville, Florida could be predicted as,

$$Y = -107.5 + 365.7 \log [(x_1/6.9) + 6] - 4560 [\log (x_2x_3/645)]/(80 + x_1/25.4) - 1.438 (x_4 + x_5 + x_6 + x_7) \quad (1)$$

where:

Y = dry weight density (kg/m^3)
x_1 = applied pressure (kPa)
x_2 = mean particle size (mm)
x_3 = standard deviation of the particle size distribution (mm)
x_4 = percent composition of garden waste
x_5 = percent composition of cardboard
x_6 = percent composition of paper
x_7 = percent composition of wood

This equation followed the general pattern characteristic of most soils, in that the density is primarily dependent upon the logarithm of the applied pressure. Although Ruf's model can be used to model the behavior of RDF in the early stage of compression, it is of limited value in modeling a pellet-forming process. The reason for the limitation is that pellet formation generally occurs at pressures of about an order of magnitude greater than those used by Ruf.

Attempts to characterize the pelletization and cubing of RDF have been made by the National Center for Resource Recovery (NCRR) the

University of California at Berkeley (UCB), and Cal Recovery Systems, Inc. (CRS). Much of the work has been sponsored by the U.S. Environmental Protection Agency (EPA). Hainsworth, et al [5], in summarizing the results of tests at NCRR, present the following equation for predicting the power required to pelletize RDF in a rotary die and roller type pellet mill:

$$PC = 55.6\ m + 21.8 \tag{2}$$

where:

PC = power consumption (kW)
m = dry mass throughput rate (Mg/hr)

This equation is in severe conflict with other data presented by NCRR.

In a study conducted at the NCRR [6,7], it was found that the specific energy consumption by the pellet mill used in the study dropped from 18 kWh/Mg to about 4 kWh/Mg as the mass throughput was increased from 1.4 Mg/hr to 8.2 Mg/hr. At mid-range, the energy requirement was 6 to 7 kWh/Mg. The NCRR investigation involved the use of screened and air-classified refuse. The particle size was less than 19 mm. The pellet mill was rated at 9 Mg/hr and was powered by a 110 kW motor. Two sizes of rollers were tried. To pelletize RDF at a given flowrate, less energy was required with the 330 mm diameter rollers than with the 250 mm rollers. Dies of two sizes were used, namely, 13 and 25 mm. A greater mass throughput rate was possible with the larger die. A significant observation was that although it required more energy to pelletize the drier material, the resulting pellets were denser and more durable than was the case with the moister material.

The cost of densifying RDF was estimated to be about $2 per Mg in 1977. This cost included a maintenance expenditure of 10 percent of the capital cost. The capital cost was about $400,000 for a 23 Mg/hr (output) plant equipped with three pellet mills. Wiles [8], in reviewing the NCRR results, noted that the useful lives of the rollers and dies used at the NCRR were 450 Mg and 900 Mg, respectively. He predicted that their life-spans could be trebled by lowering the concentration of inert matter in the RDF to 8 to 12 percent. He further noted that the expensive rollers and dies could be reconditioned instead of being replaced.

In their work at the NCRR, Arnold and Hendrix [9] found that 49 percent of the incoming raw wastes ended up in the RDF stream, inasmuch as the remaining 51 percent had been removed in the air classification and screening steps. The amount of mass lost in the pelletizing steps is insignificant, whereas the density is increased by a factor of 20.

Results of research at the University of California at Berkeley (UCB) [10] with the use of a 56 kW pellet mill rated at 2 Mg/hr throughput for animal feed, showed that only 0.6 to 1.0 Mg of RDF could be densified per hour. The machine could not process the material at a higher feed rate, although the power consumption never exceeded 25 kW. When screened light fraction was used as the feedstock, the specific energy consumption ranged from 26 kWh/Mg to 133 kWh/Mg with a 19 x 76 mm die and 42 kWh/Mg to 190 kWh/Mg with a 25 x 127 mm die as the mass throughput rate was increased from 0.1 to 0.6 Mg/hr. When the air classified light fraction was densified, the specific energy consumption was lower at all feed rates.

The specific energy consumption was found to fit the equation,

$$E = A\,\dot{m}^{b} \tag{3}$$

where:

E = Specific energy (kWh/Mg)

$\dot{m}$ = mass throughput rate (Mg/hr)

A and b are parameters that depend on the die and material.

Researchers at the UCB also used a single-die laboratory apparatus to investigate the process of pellet formation and flow through a die [10]. In the single-die study, the following three distinct phenomena were identified: (i) pre-compression in which loose RDF is compacted; (ii) material deformation in which compacted material is deformed to fit the shape and diameter of the die; and (iii) friction between the die wall and the pellet being extruded. The first phenomenon is similar to that studied by Ruf [4], in that the pressure varies exponentially with the density. Material deformation was identified as the main contributor to the energy consumption in the apparatus used. The apparatus deformed material from an initial diameter of 25 mm to a final diameter of 13 to 19 mm. Friction appeared to conform to theory, according to which

$$P = P_0 \exp [4\mu L/D] \quad (4)$$

where:

P = the pressure to overcome friction

P_0 = a constant

μ = coefficient of kinetic friction

L = length of die

D = diameter of die

For conditions typically encountered with RDF, the constant P_0 was found to be about 1000 kPa; and the coefficient of friction, about 0.1. It was noticed, however, that in practice, a pellet stops and starts repeatedly while passing through a die. The coefficient of static friction, typically about 0.16, may significantly contribute to the force opposing pellet motion.

Other trends identified in the study were:

1. Increasing the moisture content by 1 percent resulted in a drop in pressure requirement by about 200 kPa.
2. Temperature within the range of 70 to 200°F had no apparent effect on energy requirements.
3. An increase in the fraction of newsprint in the feedstock caused a dramatic rise in the pressure requirement.

In evaluating the utility of a Papakube densifier for the U.S. Navy, CRS personnel [11] found that the typical specific energy requirements were 8 kWh/Mg with 32 x 32 mm dies, and 9 kWh/Mg with 25 x 25 mm dies. The respective feed rates were 6 to 12 Mg/hr and 3 to 10 Mg/hr. The energy for densification was about one-fourth the total energy required for processing Navy waste (from raw waste to d-RDF).

Reed and Bryant [12] of the Solar Energy Research Institute (SERI) summarized the state-of-the-art of biomass densification in 1978. They cited data supplied by California Pellet Mill, Inc., which indicated that at a production rate of 10 Mg/hr, about 15 kWh are consumed per Mg of MSW pelletized. About twice as much energy is required to densify wood. The SERI report included an analysis of the cost of adding a pelleting operation to an existing 300 Mg/hr (input) refuse processing plant. The break-even selling price of the d-RDF was $21.60/Mg. The break-even price was highly sensitive to local dumping costs and the plant capacity factor.

5.1 REFERENCES

1. Bellinger, P.L., and H.F. McColly, "Energy Requirements for Forming Hay Pellets," Agricultural Engineering, 42(4):180-181; 42(5):244-247, 250 (1961).

2. Pfost, H.B., "The Effect of Lignin Binder, Die Thickness and Temperature on the Pelletizing Process," Feedstuffs, 36(22):20-21 (1964).

3. Dobie, J.B., "Cubing Tests with Grass Forages and Similar Roughage Sources," Transactions of the ASAE, 18(5):864-866 (1975).

4. Ruf, J.A., "Particle Size Spectrum and Compressibility of Raw and Shredded Municipal Solid Waste," Ph.D. Thesis, University of Florida (1974).

5. Hainsworth, E., J.L. Mayberry, and R.R. Piscitella (EG&G Idaho, Inc.), "Energy from Municipal Solid Waste - Mechanical Equipment Status Report (Draft)," U.S. Department of Energy, Contract No. DE-AC07-761D01570 (1982).

6. Renard, M.L., Refuse-Derived Fuel (RDF) and Densified Refuse-Derived Fuel (d-RDF), National Center for Resource Recovery, No. RM77-2 (1978).

7. National Center for Resource Recovery, Summary of Project and Finding -- Preparation of d-RDF on a Pilot Scale (1977).

8. Wiles, C.C., "The Production and Use of Densified Refuse Derived Fuel," Fifth Annual Research Symposium, Land Disposal and Resource Recovery, U.S. Environmental Protection Agency, Office of Research and Development, Orlando, Florida (1979).

9. Arnold, J.M. and D.C. Hendrix, Initial Mass Balance for Production of Densified Refuse-Derived Fuel, National Center for Resource Recovery, No. RR77-3 (1977).

10. Trezek, G.J., G.M. Savage, and D.B. Jones (University of California), Fundamental Considerations for Preparing Densified Refuse Derived Fuel, U.S. Environmental Protection Agency, Grant No. R-805414-010 (1981).

11. Cal Recovery Systems, Inc., Conversion of Navy Waste to d-RDF by the Papakube Process and Identification of Commercial Sources, for U.S. Navy, Civil Engineering Laboratory (1979).

12. Reed, T. and B. Bryant (SERI), Densified Biomass: A New Form of Solid Fuel, for U.S. Department of Energy, Contract No. EG-77-C-01-4042 (1978).

6. Ferrous Separation

The recovery of metals through the use of magnetic separation is considered to be one of the more effective and economical, as well as the simplest, of the unit processes employed in the resource recovery field. Relative to other metals, iron and steel alloys have a very strong magnetic permeability. Therefore, magnetic metals can be removed from a mixture of materials by passing the materials through a magnetic field [1,3,10,11]. There are two general applications of magnetic separation: (1) removal of unwanted magnetic metals from a feed stream; and (2) concentration of magnetic materials for reuse. Magnetic separation is a well-established process that has been widely used for many years to concentrate iron ores and remove tramp iron from scrap [1,3,12,13].

The two major types of magnetic separators used in the solid waste management industry are drum magnets and overhead belt magnets [4]. The advantages and disadvantages of one type versus the other one have been well discussed in articles by Handler and Runyon [6], Tobert [7], Alter, et al [8], and Twichell [9].

With specific reference to the modeling of magnetic separation processes, Trezek, Diaz, and Savage developed a first-order model of the process using the concept of a transfer function and an average recovery efficiency of 90 percent [14].

Typically, the magnetic drum is installed over the discharge end of a conveyor or feeder or is suspended above the head pulley. Overhead belt magnets usually are installed over conveyors that are used to transport shredded refuse [4]. The actual location of the unit depends upon the design of the plant. Typically, magnetic separation takes place directly after the raw refuse: (1) has been trommelled; (2) has been sized reduced; or (3) has been air classified [2,4,11].

The theory of magnetic separation has been discussed and the parameters that affect the degree of separation have been described in several papers and publications, among which are Refs. 11, 12, and 13. The ability of a magnet to attract magnetic metals is affected chiefly by: (1)

the flux density; and (2) the rate of change of the flux density. Other factors that must be considered in designing a magnetic metal separation system are object size; distance between magnetic separator and object; depth of burden; and the capacity, width, and speed of the conveyor belt [11,12].

Recovery efficiencies as low as 32 percent [2] and as high as 87 percent [8] have been reported. System costs have been reported as being as high as $1920/Mg/hr [4] and as low as $1153/Mg/hr [2]. Simister and Bendersky [2] report the energy consumption in magnetic separation as being as low as 0.3 kWh/Mg, whereas in a NCRR Bulletin [5] a consumption of 2.3 kWh/Mg is stated.

6.1 REFERENCES

1. Alter, H. and K. L. Woodruff, Magnetic Separation: Recovery of Salable Iron and Steel from Municipal Solid Waste, U.S. EPA, Report SW-559, pp. 24 (1977).

2. Simister, B.W., and D. Bendersky, Processing Equipment for Resource Recovery Systems, Volume II, Magnetic Separators, Air Classifier, and Ambient Air Emissions Tests, U.S. EPA-600/2-80-007b, pp. 160 (July 1980).

3. Alter, H., S.L. Natof, K.L. Woodruff, and R.D. Hagen, The Recovery of Magnetic Metals from Municipal Solid Waste, National Center for Resource Recovery, Inc., Report RM 77-1, pp. 59 (November 1977).

4. NCRR, "Magnetic Metals Recovery -- A Review," NCRR Bulletin, pp. 101-107 (Fall 1977).

5. NCRR, "Magnetic Metals Recovery -- A Review," NCRR Bulletin, pp. 18-27 (Winter 1978).

6. Handler, I., and K. Runyon, "Performance and Testing of the Ferrous Metals Recovery System at Recovery 1," in Proceedings of the 1980 National Waste Processing Conference, Washington, D.C., pp. 173-188 (May 1980).

7. Tobert, R., "Belt Type Magnets on Drum Magnets which Best Serve Resource Recovery," Solid Wastes Management, 19(2), (February 1976).

8. Alter, H., J. Arnold, R.J. Lotito, and S.L. Natof, "Probing the Attraction of Ferrous Recovery via Magnets," Waste Age, 33-34 (January 1979).

9. Twichell, E.S., "Our Company's Approach to Ferrous Extraction," Solid Wastes Management, 18(11), (November 1975).

10. Vesilind, P.A., and A.E. Rimer, Unit Operations in Resource Recovery Engineering, Prentice Hall, Inc., New Jersey, pp. 452 (1980).

11. Diaz, L.F., G.M. Savage, and C.G. Golueke, Resource Recovery from Municipal Solid Wastes, Volumes I and II, CRC Press, Inc. (1982).

12. Gunther, C.G., Electro-Magnetic Ore Separation, McGraw Hill Book Company, New York, NY (1909).

13. Gaudin, A.M., Principles of Mineral Dressing, McGraw Hill Book Company, New York, NY (1939).

14. Trezek, G.J., L.F. Diaz, and G.M. Savage, Prediction of the Impact of Screening on Refuse-Derived Fuel Quality, under contract to Electric Power Research Institute, EPRI FRP-1249 (November 1979).

7. Non-Ferrous Separation

Generally, the mechanical recovery of aluminum from mixed municipal solid waste requires a series of processing steps designed to size reduce and produce a fraction that has a high concentration of aluminum. Typically, size reduction is combined with one or more steps of air classification and magnetic separation in order to produce a relatively pure concentrate. The concentrate serves as a feedstock to an aluminum recovery device [1].

The technology for recovering aluminum from MSW has evolved into the following three basic techniques or systems: (1) heavy (dense) media; (2) electrostatic; and (3) eddy current separation. Efforts to model aluminum recovery systems in a resource recovery facility have not been reported in the literature.

7.1 HEAVY MEDIA

In this system, separation is accomplished through the use of fluids that have specific gravities greater than that of water. The specific gravity of the fluid usually is adjusted through the addition of colloidal solids. In this manner, specific gravities greater than 3.0 can be obtained. Aluminum floats in material that has a specific gravity of about 2.6 and sinks in material that has a specific gravity of approximately 1.4 [2,3]. Generally, heavy media separation is preceded by water elutriation. In water elutriation, water is forced upwards through the material to be separated in order to produce an apparent specific gravity of between 1.0 and 2.0. Water elutriation has been found to be effective in the removal of light, organic materials. The removal of organic materials is very important, because its presence is conducive to the loss of heavy media [4,5,6].

Historically, dense media separation has been used in the treatment of mining residues. To be efficient and cost-effective, the process generally requires throughputs on the order of 10 to 25 Mg/hr [6].

The economic feasibility of using heavy media separation for reclaiming aluminum from MSW is not clear. Michaels, _et al_ [5] projected a reasonable performance for a facility processing 50 to 100 Mg/hr of MSW. At that time (1975), the authors estimated a capital cost of $300,000 and an operating cost of $0.3/Mg of MSW processed. Abert [2] states that if it is done on a small scale, the initial cost of a heavy media system is high.

Another approach in dense media separation takes advantage of the properties exhibited by magnetic fluids. A magnetic fluid consists of sub-micronic particles of a magnetic material suspended in a hydrocarbon liquid base, such as kerosene or heptane. The imposition of a magnetic field on the fluid can be used to adjust the apparent fluid density from 1.0 to about 20.0. Thus, separation of materials on the basis of density can be achieved [6,14].

7.2 ELECTROSTATIC SEPARATION

Electrostatic separation is based upon the principle that charged particles exposed to electrostatic forces respond to laws of attraction and repulsion (Coulomb's Law). Thus, separation can be carried out by electrically charging the material to be separated, and then attracting it or repelling it with electrodes of opposite or like charge.

Experimental data indicate that 78 to 84 percent of the aluminum can be recovered from the heavy fraction of an air classifier by means of electrostatic separation [15].

7.3 EDDY CURRENT SEPARATION

Changes in the magnetic induction in a material result in the generation of a voltage in that material. In an electrical conductor, the induced voltage generates current loops called eddies. The direction of the current loops is a function of the intensity of the magnetic flux applied. If the flux is increasing, the direction of the loops is such that a magnetic field is generated that opposes the applied magnetic field. On the other hand, if the flux is decreasing, the direction of the eddy current generates a field that strengthens the applied field.

Four methods have been described for inducing eddy currents in metals. They are: (1) physically moving the material through a magnetic field; (2) moving a magnetic field through the material by electrical phasing techniques; (3) moving a magnetic field through the material by physically moving the magnet; and (4) temporarily changing the intensity of the magnetic field in the material [16].

Studies have demonstrated that eddy current separation can be used to recover aluminum from MSW. Among the key parameters that affect the efficiency of the process are electrical resistivity, size and shape of the aluminum items, and magnitude of the induced force [2,7,8,9,10,12,13].

There are three general types of aluminum recovery systems based on the principle of eddy current separation [2,7,8,9,10]. In two of the systems, linear induction motors are used. In the third system, separation is accomplished through the use of a permanent magnetic field.

In one system, material is fed onto a non-magnetic belt. The belt travels over linear induction motors positioned beneath the belt. Aluminum is forced to one edge of the belt, whereas the path of the rejects remains unaffected. The extent of aluminum recovery with this system is relatively great, especially after two passes [2,7].

In another system, four magnets are used. Two of the magnets are placed above and two below the conveyor belt. The unit operates such that the first two magnets force the aluminum to one side of the belt, and the second pair of magnets forces the material off to the opposite edge. About 75 to 80 percent of the recovered fraction is aluminum. The remainder consists of other non-ferrous metals (10 to 15 percent) and organic matter (10 percent) [17]. Extent of aluminum recovery by means of this unit is a function of feed rate.

Vertical eddy current separators have also been used. In such systems, the material to be segregated is allowed to fall through two parallel banks of linear induction motors. As the material falls, the motors are operated such that the magnetic field is moved upwards. As a result, the metal is also forced upwards, thereby allowing its separation. This system requires about 2 to 10 kW per Mg/hr. Results of experiments show that 80 to 95 percent of the aluminum cans can be recovered [2]. The efficiency of the system depends upon size of the aluminum particles and feed rate [13].

A system to recover non-ferrous metals through the use of permanent magnets has been developed [9,1]. The system has no moving parts and consumes no power. Basically, the unit consists of an inclined ramp in which permanent magnet strips of alternating polarity are embedded. The strips are placed at a 45° angle to the ramp's axis. The material to be segregated is fed at the top of the ramp and is allowed to slide down the ramp. Non-metallic particles are not affected by the magnet strips and slide down unimpeded. The metallic particles, however, are deflected due to the fact that they are moving through a magnetic field. Experimental work indicates that a separator processing about 1 Mg of material per hour can achieve a non-ferrous metal recovery efficiency greater than 70 percent; and that the purity would exceed 80 percent [9]. Key variables that affect separation efficiency are electrical conductivity, size and shape of particle, strength of magnetic field, coefficient of friction between particles and ramp, and length and inclination of the ramp [9,18].

7.4 REFERENCES

1. Dean, K.C., E.G. Valdez, and J.H. Bilbrey, Jr., "Recovery of Aluminum from Shredded Municipal and Automotive Wastes," Resource Recovery and Conservation, 1:55-56 (1975).

2. Abert, J.G., "Aluminum Recovery -- A Status Report," NCRR Bulletin, VII:(2) (Spring 1977).

3. Vesilind, P.A., and A.E. Rimer, Unit Operations in Resource Recovery Engineering, Prentice-Hall, Inc., New Jersey (1981).

4. Testin, R.F., "Recovery of Aluminum from Solid Waste," Presented at Stanford Research Institute Symposium on Solid Wastes (October 1974).

5. Michaels, E.L., et al, "Heavy Media Separation of Aluminum from Municipal Solid Waste," Transactions of ASME, 258:34-39 (March 1975).

6. Testin, R.F., "Recovery of Non-Ferrous Metals from Solid Wastes," Presented at the American Institute of Chemical Engineers Symposium on Recycling, Atlantic City, New Jersey (1976).

7. Morey, B., and S. Rudy, "Aluminum Recovery from Municipal Trash by Linear Induction Motors," Presented at 103rd Annual Meeting of the American Institute of Mining, Metallurgical, and Petroleum Engineers, Dallas (February 1974).

8. Campbell, J.A., "Electromagnetic Separation of Aluminum and Nonferrous Metals," Presented at 103rd Annual Meeting of the American Institute of Mining, Metallurgical, and Petroleum Engineers, Dallas (February 1974).

9. Spencer, D.B., and E. Schloemann, "Recovery of Non-Ferrous Metals by Means of Permanent Magnets," Waste Age, 6(10):32 (October 1975).

10. Schloemann, E., "A Rotary Drum Metal Separator Using Permanent Magnets," Resource Recovery and Conservation, 2:147-158 (1976).

11. Weismantel, G.E., "Needs and Know How Boost Aluminum Recycle," Chemical Engineering (May 1977).

12. Abert, J.G., "Aluminum Recovery -- A Status Report," NCRR Bulletin, VII:(3) (Summer 1977).

13. Schloemann, E., "Separation of Nonmagnetic Metals from Solid Waste by Permanent Magnets - II. Experiments on Circular Disks," Journal of Applied Physics, 46:(11) (November 1975).

14. Rosenweig, R.E., "Material Separation Using Ferromagnetic Liquid Techniques," J. AIAA, 3:(483)969 (December 1969).

15. Knoll, F.S., "Recovery of Aluminum by High Tension Separation," AIME, Preprint 74-B-28, AIME Annual Meeting, Dallas (1974).

16. Sommer, J. Jr., and R. Kenny, "An Electromagnetic System for Dry Recovery of Non-Ferrous Metals from Shredded Municipal Solid Wastes," Proceedings of 4th Mineral Waste Utilization Symposium, ITT Research Institute, Chicago, 1975, pp. 78-84.

17. Alter, H., S. Natof, and L.C. Blayden, "Pilot Studies, Processing MSW and Recovery of Aluminum Using an Eddy Current Separator," 5th Mineral Wastes Utilization Symposium, ITT Research Institute, Chicago, 1976, pp. 161-168.

18. Spencer, D.B., and E. Schloemann, "Recovery of Non-Ferrous Metals by Means of Permanent Magnets," Resource Recovery and Conservation, 1:151-165 (1975).

8. Glass Separation

The literature on resource recovery includes several publications that deal with the various systems proposed, tried, or presently in use for separating glass from the municipal waste stream. However, any effort to model a glass sorting system in a resource recovery operation has not as yet been reported in the literature.

In a resource recovery operation, glass separation usually (but not necessarily) is preceded by certain unit processes, among which are size and density classification and magnetic separation. For certain types of glass removal systems (froth flotation), it may be necessary to size reduce the waste in advance of glass removal. Because a glass separation system is comprised of a number of unit processes, the important variables can be identified only by considering each unit process separately. If froth flotation is the method of separation, then size reduction is added to the unit processes that precede the separation step.

8.1 FROTH FLOTATION

At the start of this section, it should be pointed out that most of the information in the literature on the use of froth flotation systems to recover glass is confined to process descriptions. Very few data are given on the economics of the process and or mass and energy balances.

When glass separation is to be accomplished by froth flotation, the flotation step is preceded by two steps or operations that are designed to provide a glass-rich slurry. Named in the order of their application, they are "jigging" and "slime" (fines) separation by a hydrocyclone. A jig is used to separate the lighter particles from the heavier particles in a mixture. Separation is a function of the differences between the tendencies of the various types of particles to penetrate a pulsating bed -- in this case, water. Vesilind and Rimer [1] state that the important material variables are the specific gravities of the components and to a lesser extent, the sizes and shapes of the particles. Machine variables are sieve size and the magnitude and frequency of the pulsating water forces. Taggart [2] has developed a formula for the power consumption of a jig in terms of sieve area and feed particle diameter.

In the sequence, the function of the hydrocyclone is to separate particles smaller than a selected size (i.e., 150 to 200 mesh) from the slurry, which as a result of the jigging, now consists mostly of glass particles. According to Vesilind and Rimer [1], the most important variables in the hydrocyclone process are the specific gravity and size distribution of the solids particles, viscosity of the liquid medium, solids content of the slurry, and rate of flow through. Important machine variables are the diameter of the cyclone and the magnitude of the pressure drop.

McChesney and Degner [3] describe an approach to the recovery of glass that involves the removal of organics by hydraulic separation, ferrous metal by magnetic separation, and heavy non-ferrous metals by heavy media separation. The material remaining after the three separations is crushed, screened, and then subjected to froth flotation. Possible feeds to the glass recovery system are the air classified heavy fraction of refuse and incinerator residue.

In the course of their article, McChesney and Degner discuss the principles of froth flotation. According to the two researchers, important process variables are rotor speed, rotor submergence, "pulp" conditioning, and residence time. ("Pulp" is industry jargon for pulverized glass.) The term "conditioning" refers to the adding of reagents to the froth flotation feed. The additions affect the surfaces of the glass particles such that they become hydrophobic. Because of the change in surface characteristic, the glass particles adhere to air bubbles generated in the liquid in the flotation cell and are thus carried by the bubbles to the surface and become a part of the froth that is formed. The froth with its burden of glass is skimmed off and suitably treated. The heavy inerts that are not glass sink to the bottom of the vessel.

McChesney and Degner state that separation is most complete when the glass particles in the flotation feed have a size distribution of -28 to +150 mesh. Solids content of the feed should be in the range of 25 to 35 percent. McChesney and Degner maintain that if properly done, froth flotation can result in a product that is 99 percent glass.

Gershman [4] indicates various flowrates in a process diagram of a glass recovery operation in which froth flotation is used. Duckett [5] has described the froth flotation process in use at the NCRR Equipment Test and Evaluation Facility.

The froth flotation research conducted by the Bureau of Mines is the subject of a paper by Heginbotham [6]. Heginbotham describes batch flotation tests that were carried out to evaluate the effectiveness of various cationic collectors (i.e., pulp conditioners). As a part of the description, he mentions the cost of the collectors. He reports the results of the operation of a pilot plant in which 4 Mg of glass-rich aggregate were obtained from urban refuse.

Morey and Cummings [7] describe the Garrett glass recovery process, a system that is based upon froth flotation. In their paper are presented estimates of what the incremental capital and operating costs of a froth flotation glass recovery system would be if it were a part of a 2000-Mg/day pyrolysis plant. Capital costs would amount to $452,000, and operating costs would be $6/Mg of glass recovered (1972 dollars).

8.2 OPTICAL SORTING

As its name implies, optical sorting is based upon the utilization of a photocell to distinguish glass from nonglass. Significant material variables in optical sorting are particle size and light transmission characteristics. The optimum particle size is rather large, namely, 6 to 19 mm.

In his publication, Berghman [4] also briefly describes a demonstration plant at Franklin, Ohio, in which optical sorting was used. However, the test facility is described more completely in a publication by Garbe [8]. The material that is processed in the test facility at Franklin, Ohio, is the heavy fraction from a liquid cyclone. The source of the heavy fraction is refuse that has been hydropulped and passed through a magnetic separation system. After mechanical dewatering, particles less than 6 mm in size are removed by means of a vibrating screen. The retained material is subjected to magnetic and heavy media separation, jigging (to remove aluminum), drying, and electrostatic separation (to remove any remaining metals). The dried material remaining after all of the preceding separatory steps is ready for photocell sorting.

The first step in the optical sorting process is based upon the transparency of glass. In this step, glass is separated from ceramics and rocks. The second step is based upon differences in color. In this step, glass is sorted into three colors, namely, amber, green, and flint (no color). Tests conducted for the U.S. EPA indicated that slightly more than 60 percent of the glass fed into the recovery system could be recovered [9]. According to Garbe [8], the projected capital and operating costs of a glass recovery sub-system in a 1000 Mg/day facility would be respectively $2,430,000 and $1.23/Mg of MSW.

Optical sorting is carried on at the resource recovery facility in Doncaster, England [1]. In the facility, the glass separation circuit receives raw refuse (0.5 to 1.5 in. particle size) and screened, air classified heavy fraction that has been exposed to magnetic separation. Organics and non-ferrous materials are removed with the use of modified stoners followed by a rising-current separator and dewatering. The material thus treated is subjected to optical sorting.

8.3 REFERENCES

1. Vesilind, P.A. and A.E. Rimer, Unit Operations in Resource Recovery Engineering, Prentice-Hall International, New Jersey (1981).

2. Taggart, A.F., Handbook of Mineral Dressing, John Wiley and Sons, Inc., New York (1945).

3. McChesney, R.D., and V.R. Degner, "Methods for Recovering Metals and Glass," Pollution Engineering, 7(8), (1975).

4. Gershman, H.W., "Status Report on Glass Recovery," The Glass Industry, p. 24 (October 1976).

5. Duckett, J., "Glass Recovery and Reuse," NCRR Bulletin, VIII(4), (Fall 1978).

6. Heginbotham, J.H., "Recovery of Glass from Urban Refuse by Froth Flotation," Proceedings of the 6th Mineral Waste Symposium, U.S. Bureau of Mines, IIT (May 1978).

7. Morey, B., and J.P. Cummings, "Glass Recovery from Municipal Trash by Froth Flotation," Proceedings of the 3rd Mineral Waste Utilization Symposium, U.S. Bureau of Mines, IIT (March 1972).

8. Garbe, Y., "Color Sorting Waste Glass at Franklin, Ohio," Waste Age, p. 70 (September 1976).

9. Systems Technology Corporation, A Technical, Environmental and Economic Evaluation of the Glass Recovery Plant at Franklin, Ohio, EPA Report No. SW-146c (1977).

9. Conveying

Design information and experimental data are available in the literature on conveying. However, nothing has been published on the modeling of a conveying process.

The Conveyor Equipment Manufacturers Association [1] has published a comprehensive design manual for belt conveyors. In the manual, it is pointed out that the important properties and characteristics of bulk materials in terms of conveyability are size, bulk density, angle of repose, and abrasiveness. Belt width, speed, and angle of inclination are among the important operational parameters listed. Methods for determining capacity and power requirement are given.

In a chapter written for the Handbook of Mineral Dressing [2], Behre described the conveyors utilized in the mining industry. His description covers the belt, pan, apron, flight, bucket, and screw conveyor types. In addition to a discussion of the important material properties and operational variables, the chapter also presents empirical formulas that can be used to determine conveyor capacity and power requirements. Because the Handbook was published so long ago (i.e., 1945), the cost of belt conveyor systems presented in the chapter must be taken in that context.

Experimental work involving conveying of processed refuse fractions has been carried out and reported by Khan, *et al* [3]. Properties of the materials conveyed including bulk density, angle of repose, and angle of maximum inclination were measured; and belt, vibrating pan, and apron conveyors were tested. The rate of spillage from the belt and vibrating pan conveyors was measured at a variety of mass flowrates and belt velocities. In tests with the apron conveyor, maximum carrying capacity was measured as a function of conveyor velocity and inclination. In the same research program, a plan for testing a pneumatic conveying system for transporting solid waste fractions was developed [4]. Principal operational parameters listed are air velocity, solids-to-air ratio, and duct diameter. Material size and shape, composition, moisture content, bulk density, angle of repose, and abrasiveness are the key material properties and characteristics indicated. A method for determining the total system pressure drop is presented. The article includes information on the cost and sizing of a conveying test.

9.1 REFERENCES

1. Conveyor Equipment Manufacturers Association, Belt Conveyors for Bulk Materials, CBI Publishing Co., Inc., Boston, Mass. (1979).

2. Behre, H.A., "Transport of Materials," from Handbook of Mineral Dressing by A.F. Taggart, John Wiley and Sons, Inc., New York, NY (1945).

3. Khan, Z., M.L. Renard, and J. Campbell, Considerations in Selecting Conveyors for Solid Waste Applications, National Center for Resource Recovery, Inc., Washington, D.C., EPA-600/2-82-082 (September 1982).

4. Renard, M.L., A Pneumatic Conveying Test Rig for Municipal Solid Waste Fractions, National Center for Resource Recovery, Inc., Washington, D.C., EPA-600/2-82-083 (September 1982).

Part II

Task 3 Report: Mass Balance and Energy Requirement Models

Preface

The Task 3 Report presents the models for the mass balance and energy requirements for the following generic unit operations:

- Size Reduction
- Air Classification
- Trommel Screening
- Ferrous Separation
- Densification
- Conveying

Also presented in the report are the derivations of the models and their limitations. In those cases where test data are available, calculations are presented to illustrate the use of the models and their accuracy.

1. Size Reduction Model

1.1 BACKGROUND

The size reduction of solid waste is employed primarily for three reasons in the field of resource recovery. First, shredders reduce the size of large items so that the material can be handled efficiently by conventional processing and material handling equipment. Secondly, shredders liberate and expose materials so that they can be separated and recovered. Lastly, in those cases where final particle size is of importance (e.g., RDF preparation), size reduction can be employed to produce the required top size.

The size reduction model is formulated to describe the product size distribution and energy requirements associated with solid waste comminution under given conditions of feed size distribution and of machine configuration. The model is structured to simulate refuse comminution as a function of the type of components comprising the feedstock. The size reduction model for municipal solid waste has been developed to simulate the performance of horizontal hammermills. Horizontal mills were chosen for several reasons. First, they represent the major category of equipment used in refuse size reduction. Secondly, hammermills have been the subject of previous modeling efforts in the area of refuse comminution [1,2,3]. Lastly, there exists in the case of horizontal hammermills a substantial amount of field test data.

The basic block diagram for the hammermill shredder model is shown in Figure 1.1.

1.2 DEVELOPMENT OF MASS BALANCE MODEL

1.2.1 Description of Model

The solid waste size reduction model has been developed using the concept of linear, size-discrete comminution kinetics. The approach has been followed previously by a number of researchers in the field of mineral comminution. With regard to the development of a model for the size

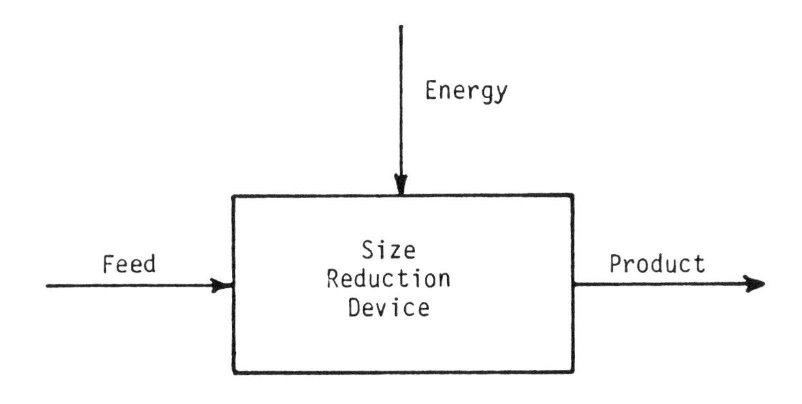

Figure 1.1. Block Diagram of the Refuse Size Reduction Model

reduction of solid waste, the linear, size-discrete approach allows convenient matrix representation of the process. Thus the representation is amenable to simulation by digital computer. In the development of the size reduction model, the governing equations of comminution are derived initially from the consideration of the batch milling process and are subsequently extended to continuous steady-state milling by invoking the concept of the residence time distribution of material within the mill cavity.

The governing relations are developed for a single-component material and extended to multi-component size reduction through the assumption of linearity, i.e., the breakage behavior of each component is considered to be independent of the presence of other components. The model computes the size distribution for each component. The cumulative size distribution for the mixture is computed as the sum of the products of the component size distributions and their respective mass fractions present in the feed.

The model uses the concept of selection and breakage functions to describe the breakage of material within the size reduction device. The use of the functions allows different types of size reduction devices to be modeled inasmuch as both selection and breakage events are governed to a large degree by the internal geometry of the mill and by its operating conditions. In addition to the machine parameters, the properties of the material also influence its breakage within the mill.

1.2.2 Important Assumptions

The important assumptions used in formulating the size reduction model are recapped below:

1. The throughput is constant. Therefore, size reduction is accomplished under steady state conditions.
2. The breakage behavior of each component is independent of the presence of other components.
3. S_1 and B_{ij} are independent of size class and time. Therefore, the cumulative breakage function can be normalized for each size class of the feed, and the kinetic model is linear with constant coefficients.
4. The values of the cumulative breakage function are represented by the relation,

$$B_{ij} = \left(\frac{K_1}{S_1} \frac{X_i^2}{X_j X_{j+1}} \right)^{\alpha/2}$$

where:

K_1 and α are constants;

S_1 = the selection function value for the top size class; and

X = mass fraction in a given size class.

5. Residence time (τ) is related to mass throughput (Q) according to an equation of the form,

$$\tau = aQ^b$$

6. All size classes for a given component have identical residence time distributions.

1.2.3 Governing Theory

The linear, size-discrete matrix model for size reduction is formulated by dividing the feed into discrete narrow size classes. Establishing a mass balance on the material in each size interval results in the following relation [4] for open-circuit batch milling,

$$\frac{d\,(H\, m_i(t))}{dt} = -\, S_i\, H\, m_i(t) + \sum_{j=1}^{i-1} b_{ij}\, S_j\, H\, m_j(t) \tag{1}$$

where:

$m_i(t)$ = mass fraction in the ith size class;

H = total mass of material in the size reduction device at time t;

S_i = fractional rate at which material is broken out of the ith size class; and

b_{ij} = fraction of material in the jth size class that appears in the ith size class.

Invoking the assumption that S_i and b_{ij} are independent of size class and time, Eq. 1 may be rewritten in matrix notation for the complete ensemble of size classes,

$$\frac{d\,[H\, \underline{m}(t)]}{dt} = -\, [\underline{\underline{I}} - \underline{\underline{B}}]\, \underline{\underline{S}}\, H\, \underline{m}(t) \tag{2}$$

where:

$\underline{\underline{I}}$ = identity matrix;

$\underline{\underline{S}}$ = selection matrix; and

$\underline{\underline{B}}$ = breakage function matrix.

In steady-state operation, the mass of material in the size reduction device is constant. Thus, for steady-state conditions Eq. 2 becomes,

$$d \frac{[\underline{m}(t)]}{dt} = - [\underline{\underline{I}}-\underline{\underline{B}}] \, \underline{\underline{S}} \, \underline{m}(t) \tag{3}$$

for which the analytical solution is,

$$\underline{m}_b(t) = \exp \left[-(\underline{\underline{I}}-\underline{\underline{B}}) \, \underline{\underline{S}} \, t\right] m_b(o) \tag{4}$$

where:

$m_b(o)$ = initial mass of material in the mill.

The subscript b denotes that the solution is for batch comminution.

In the case where no two selection functions are equal, the term $\exp [-(\underline{\underline{I}}-\underline{\underline{B}}) \, \underline{\underline{S}} \, t]$ can be simplified by a similarity transform [5]. Thus, Eq. 4 is transformed to give,

$$\underline{m}_b(t) = \underline{\underline{T}} \, \underline{\underline{J}}(t) \, \underline{\underline{T}}^{-1} \, m_b(o) \tag{5}$$

where:

$$T_{ij} = \begin{cases} 0 & i<j \\ 1 & i=j \\ \sum_{k=j}^{i-1} \dfrac{b_{ik}S_k}{S_i - S_j} T_{kJ} & i>j \end{cases}$$

$$J_{ij}(t) = \begin{cases} \exp(-S_i t) & i=j \\ 0 & i \neq j \end{cases}$$

Inasmuch as a continuous size reduction relation is sought for the refuse comminution model to be developed here, the batch comminution relation (Eq. 5) must be extended to continuous milling. The extension is made using the concept of residence time distribution [3]. The residence time distribution is defined as the mass of material of a given size that is contained within the size reduction device as a function of time. If it is assumed that a single residence time distribution characterizes all of the particle size classes, the steady-state size distribution from the mill can be represented by an average of the batch responses weighted with respect to the residence time distribution, i.e.,

$$m_{cp} = \int_0^\infty m_b(t)\ R(t)\ dt \qquad (6)$$

where:

$R(t)$ = residence time distribution; and

cp = product under continuous milling conditions.

Substituting the transformed relation for $m_b(t)$ (i.e., Eq. 5) into Eq. 6 yields,

$$\underline{m}_{cp} = \underline{\underline{T}}\ [\ \int_0^\infty \underline{\underline{J}}(t)\ R(t)\ dt]\ \underline{\underline{T}}^{-1}\ \underline{m}_{cf} \qquad (7)$$

where:

m_{cf} = mass fraction of the feed material in continuous mill operation.

The integrand in Eq. 7 is commonly expressed in terms of the dimensionless time variable, $\theta = t/\tau$, where τ is the mean resident time and is the quotient of the mass of material held within the mill and the throughput. The mean residence time τ is assumed to be related to the throughput (Q) through an equation of the form [6],

$$\tau = aQ^b \qquad (8)$$

where:

a and b are constants that represent the characteristics of the mill.

Expressing Eq. 7 in terms of θ yields,

$$\underline{m}_{mp} = \underline{\underline{T}}\ \underline{\underline{J}}_c(\tau)\ \underline{\underline{T}}^{-1}\ \underline{m}_{mf} \qquad (9)$$

where:

$$J_{c_{ij}}(\tau) = \begin{cases} \int_0^\infty R(\theta)\ \exp\ (-S_i\tau\theta)\ d\theta & i=j \\ 0 & i \neq j \end{cases}$$

To model refuse size reduction devices, the continuous open circuit relation given in Eq. 9 has been modified by the addition of an internal classifier function [4]. The classifier represents the openings (or clearance dimension) that restrict the flow of material through the mill until the size of the particles is less than the size of the openings. The steady-state description of the refuse size reduction device follows from Eq. 9 and a mass balance on the size reduction equipment [6]. A representation of the size reduction circuit is shown in Figure 1.2. Since under steady-state conditions the infeed and discharging rates are identical, the governing relation for closed circuit size reduction is

$$\underline{m}_p = [\underline{\underline{I}}-\underline{\underline{C}}]\ \underline{\underline{T}}\ \underline{\underline{J}}_c(\tau)\ \underline{\underline{T}}^{-1}\ [\underline{\underline{I}}-\underline{\underline{C}}\ \underline{\underline{T}}\ \underline{\underline{J}}_c(\tau)\ \underline{\underline{T}}^{-1}]^{-1}\ \underline{m}_f \tag{10}$$

The size discrete selection function describes the mass fraction within a discrete size class that is selected for breakage. In the refuse size reduction model a value (S_i) is chosen for the mass fraction selected for breakage in the top size class. The values of the selection function for the smaller size classes are computed from the cumulative breakage function.

The estimation of the breakage function follows from two assumptions. First, the size reduction process is linear, i.e., the values of the breakage function are independent of the size distribution of material in the mill. Secondly, the size discrete breakage function can be normalized, i.e., for material breaking into smaller size classes there is a constant ratio of breakage values that is dependent upon the ratio of the successive size intervals.

The size reduction model uses a breakage function relation presented by Epstein [7] for mineral comminution and used by Shiflett [3,6] to model refuse size reduction,

$$B_{ij} = \frac{K_1}{S_1}\left(\frac{X_i^{\,2}}{X_j\ X_{j+1}}\right)^{\alpha/2} \tag{11}$$

where:

X = size class; and

K_1 = an invariant constant.

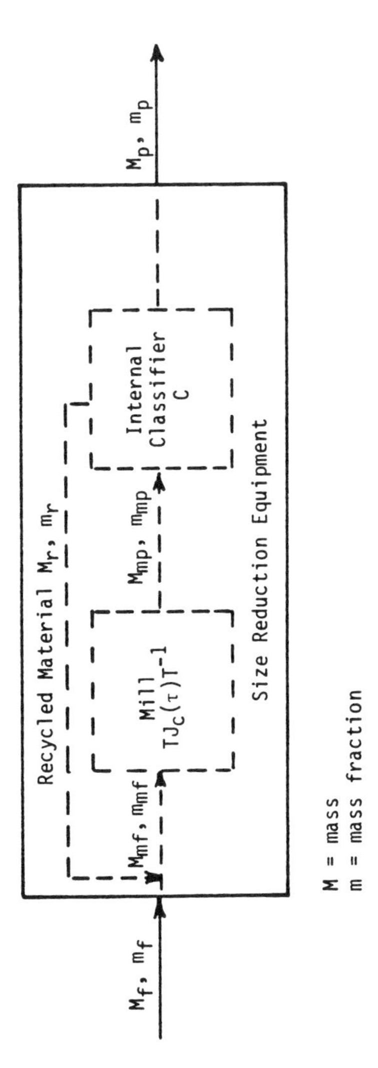

Figure 1.2. Diagram of Refuse Size Reduction Equipment with an Internal Classifier

Inasmuch as the cumulative breakage function value for the top size class is constrained by the mass balance to a value of unity, the value $K_1\ [X_i/(X_j\ X_{j+1})]^{\alpha/2}$ has been set equal to S_1 in the size reduction model for each of the top size classes. The imposition of the above constraint is a departure from the constraints imposed by Shiflett [3,6]. The constraint, however, is a necessary one in order to uphold the conservation of mass.

The computation of the cumulative breakage function values requires that α be specified. For the present model α is chosen, in addition to the S_1 value, to give an empirical fit between a set of measured product size distributions and the set of predicted values. In the case of MSW, the goodness-of-fit is constrained by the form of Eq. 11.

Based upon the limited amount of test data [8] for the size reduction of ferrous metal and newsprint, values of S_1 and α have been empirically determined. The values are reported in Table 1.1. Values of S_1 and α for other components must be estimated due to the lack of experimental data.

Table 1.1. Values of S_1 and α Chosen for Ferrous Metals and Newsprint

Component	S_1	α
Ferrous Metals	0.9	1.4
Newsprint	0.35	1.4

The following is a listing of the key inputs to the size reduction model:

- Number of Components
- Number of Size Classes
- Number of Residence Time Intervals
- MSW Composition
- Size Class Designations
- Classifier Function Values
- Residence Time/Throughput Equation Constants by Component

- Alpha (α) Values by Component
- S_1 Values by Component
- Residence Time Distribution by Component
- Component Feed Size Cumulative Percent Passing Values
- Throughput

A detailed flow chart of the size reduction model is shown in Figure 1.3.

Utilizing the available component experimental test data, a comparison of actual and predicted product size distributions has been prepared for ferrous metals and newsprint. The comparisons are presented in Figures 1.4 and 1.5 along with the pertinent input data. Entries denoted "9999" are default values used where division by zero occurs.

1.3 ENERGY REQUIREMENTS

The energy model for refuse size reduction is characterized in terms of components, similar to the development of the mass balance model. The approach is made possible due to a limited amount of component data collected previously by Savage [8] and an extension of an empirical representation of size reduction energy requirements developed previously for raw MSW and selected RDF fractions [9].

The governing relation is cast in terms of the specific energy (E_0, kWh/Mg) required to achieve a given degree of size reduction (Z). The parameter Z is expressed in terms of the characteristic size of the feed (F_0) and of the product (P_0). The values of F_0 and P_0, respectively, are numerically interpolated from the input feed size data and the product size distribution calculated by the size reduction model. The general form of the energy equation is,

$$E_0 = A\,Z^B \qquad (12)$$

where:

A and B are empirically determined coefficients; and

$Z = (F_0 - P_0)/F$.

(The characteristic size is that size corresponding to 63.2 percent cumulative passing.)

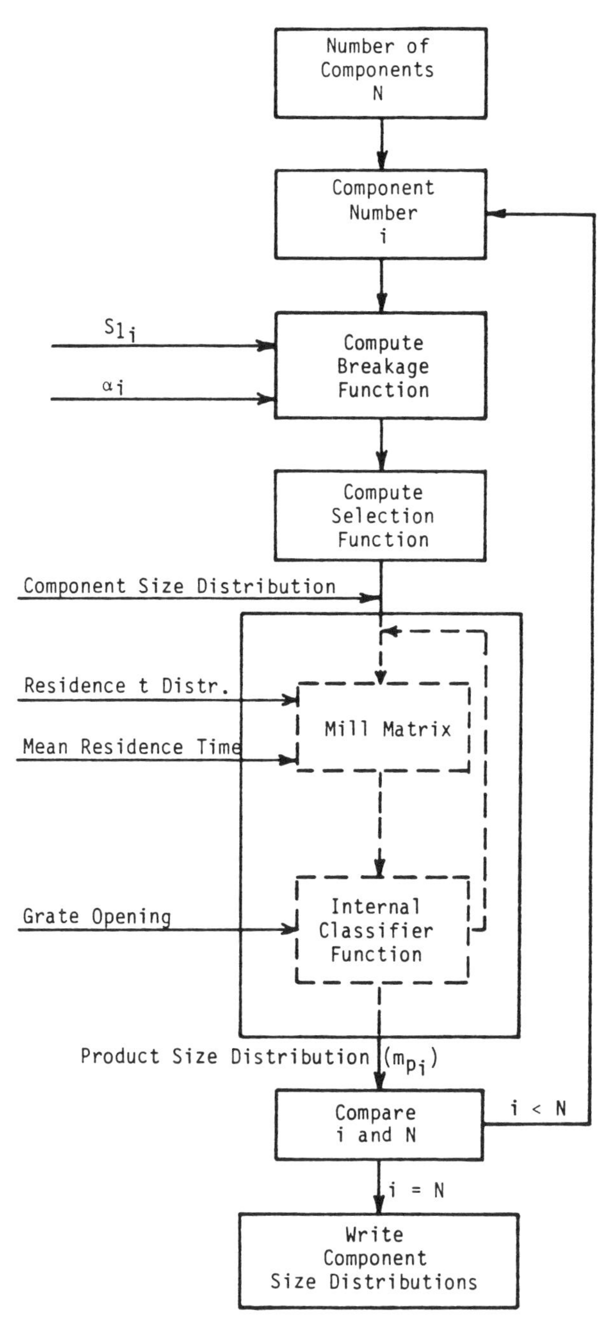

Figure 1.3. Flow Chart of Size Reduction Mass Balance Model

SIZE REDUCTION SIMULATION SUMMARY

```
COMPONENT=1                     MAX. SIZE CLASS=30.5
MSW Q (TPH)=1.84                MIN. SIZE CLASS=.0338
COMPONENT MF=.4783              K1=1.023
COMPONENT QC (TPH)=.8801
S1 = .9
GRATE OPENING=5.1               ALPHA=1.4
RESIDENCE TIME (S)=9.7582
TR=3.6*A1*QC^B1
  =3.6 (3.08) (.880072) ^ (1)
R(TH) CURVE AREA=17.76
```

	FEED INFORMATION		****************** PRODUCT INFORMATION ******************					
SCREEN	MASS	CUM.	-----MASS FRACTION------			-----CUMULATIVE % PASSING-----		
SIZE	FRACTION	PASSING	CALC'D	MEASURED	%ERROR	CALC'D	MEASURED	%ERROR
30.5	0	1	0	0	9999	1	1	0
25.4	0	1	0	0	9999	1	1	0
20.3	.068	.932	0	0	9999	1	1	0
12.7	.183	.749	0	0	9999	1	1	0
10.2	.219	.53	0	0	9999	1	1	0
5.1	.53	0	.3869	.155	-149.6185	.6131	.845	27.4448
2.5	0	0	.2127	.545	60.977	.4004	.3	-33.472
1.6	0	0	.1741	.229	23.9553	.2263	.071	-218.6952
.95	0	0	.098	.036	-172.313	.1282	.035	-266.4026
.51	0	0	.0775	.013	-495.8825	.0508	.022	-130.8009
.27	0	0	.0292	6E-03	-386.5173	.0216	.016	-34.9072
.135	0	0	.0108	.016	32.739	.0108	0	9999
.0675	0	0	5.9E-03	0	9999	4.9E-03	0	9999
.0338	0	0	4.9E-03	0	9999	0	0	9999

Figure 1.4. Size Reduction of Ferrous Metals

SIZE REDUCTION SIMULATION SUMMARY

COMPONENT=0 MAX. SIZE CLASS=30.5
MSW Q (TPH)=1.84 MIN. SIZE CLASS=.0338
COMPONENT MF=.5217 K1=.3978
COMPONENT QC (TPH)=.9599
S1 = .35
GRATE OPENING=5.1 ALPHA=1.4
RESIDENCE TIME (S)=50.1978
TR=3.6*A1*QC^B1
=3.6 (14.78) (.959928) ^ (1.424)
R(TH) CURVE AREA=17.76

FEED INFORMATION			***************** PRODUCT INFORMATION *********************					
SCREEN	MASS	CUM.	-----MASS FRACTION------			-----CUMULATIVE % PASSING-----		
SIZE	FRACTION	PASSING	CALC'D	MEASURED	%ERROR	CALC'D	MEASURED	%ERROR
30.5	.034	.966	0	0	9999	1	1	0
25.4	.366	.6	0	0	9999	1	1	0
20.3	.525	.075	0	0	9999	1	1	0
12.7	.075	0	0	0	9999	1	1	0
10.2	0	0	0	0	9999	1	1	0
5.1	0	0	.146	.039	-274.2957	.854	.961	11.1317
2.5	0	0	.3469	.315	-10.1117	.5072	.646	21.4903
1.6	0	0	.2291	.399	42.5923	.2781	.247	-12.5975
.95	0	0	.1338	.176	23.9685	.1443	.071	-103.24
.51	0	0	.0826	.049	-68.5394	.0617	.022	-180.5278
.27	0	0	.0341	.011	-210.0641	.0276	.011	-150.9915
.135	0	0	.0146	.011	-32.345	.0131	0	9999
.0675	0	0	7.3E-03	0	9999	5.8E-03	0	9999
.0338	0	0	5.8E-03	0	9999	0	0	9999

Figure 1.5. Size Reduction of Newsprint

From a limited amount of experimental data, the coefficients A and B for the case of ferrous metals and newsprint have been empirically determined. The energy equations are presented in Table 1.2. Values of

Table 1.2 Energy Requirements for Size Reduction of Selected Waste Components

Component	Specific Energy, E_o (kWh/Mg)
Ferrous Metals	$30.2\ Z_o^{1.3}$
Newsprint	$15.3\ Z_o^{1.5}$

the coefficients A and B for MSW components other than ferrous metals and newsprint cannot be verified due to a lack of experimental test data.

A comparison of the predicted and actual energy requirements for the size reduction of ferrous metals and newsprint is presented in Table 1.3 using the available test data.

Table 1.3. Comparison of Predicted and Actual Energy Requirements for the Size Reduction of Selected Components

Component	F_o (cm)	P_o (cm)	Z	Specific Energy, E_{o_i} (kWh/Mg) Predicted	Specific Energy, E_{o_i} (kWh/Mg) Actual	Percent Error
Ferrous Metals	11.3	5.2	0.5	12.3	12.0	2.4
Newsprint	25.7	3.4	0.9	12.9	13.7	6.2

Assuming linearity, the total specific energy requirement, E_o, is represented by the relation,

$$E_o = \sum_{i=1}^{n} (E_{o_i})(mf_i) \tag{13}$$

where:

n = number of components;

E_{0_i} = specific energy requirement for a given component; and

mf_i = mass fraction of component i in the feedstock.

1.4 REFERENCES

1. Obeng, D.M., Comminution of a Heterogeneous Mixture of Brittle and Non-brittle Materials, Ph.D. thesis, University of California, Berkeley, 1974.

2. Obeng, D.M., and G.J. Trezek, "Simulation of the Comminution of a Heterogeneous Mixture of Brittle and Non-brittle Materials in a Swing Hammermill," Ind. Eng. Chem. Process Des. Dev., 14(2), 1975.

3. Shiflett, G.R., and G.J. Trezek, "The Use of Residence Time and Nonlinear Optimization in Predicting Comminution Parameters in the Swing Hammermilling of Refuse," Ind. Eng. Chem. Process Des. Dev., 18(3), 1979.

4. Herbst, J.A., G.A. Grandy, and D.W. Fuerstenau, "Population Balance Models for the Design of Continuous Grinding Mills," Proc. of the Tenth IMPC, London, 1973.

5. Pease, M.C., Methods of Matrix Algebra, Academic Press, New York, 1965.

6. Shiflett, G.R., "A Model for the Swing-Hammermill Size Reduction of Residential Refuse," D. Eng. dissertation, University of California, Berkeley, 1978.

7. Epstein, B., "The Mathematical Description of Certain Breakage Mechanisms Leading to the Logarithmico-Normal Distribution, "J. Frank. Inst., 224, 1947.

8. Unpublished data collected by G. Savage, Cal Recovery Systems, Inc., on hammermilling of selected MSW components.

9. Savage, G.M., and G.J. Trezek, Significance of Size Reduction in Solid Waste Management, Volume II, EPA-600/2-80-115, Aug. 1980.

2. Air Classification Model

2.1 BACKGROUND

Air classification can be used to separate organic and inorganic materials in the processing of refuse as a feedstock for fuel preparation, fiber recovery, or composting. Air classification separates the heavy items in the waste stream, such as glass, metals, rock, leather, rubber, and dense plastics from the light items in the waste stream, such as paper and plastic. Some organic components with a high moisture content, such as yard waste, food waste, and wet paper, also typically report to the heavy fraction.

Air classification involves the passing of an air stream through the refuse stream in an enclosed chamber. Usually, the refuse is shredded prior to the air classification stage. Separation of particles is achieved as the result of the interplay of drag forces and gravitational forces. Particles with a high drag-to-weight ratio are carried away in the "light fraction," and particles with a low drag-to-weight ratio fall to the "heavy fraction." A simplified block diagram of an air classification process is presented in Figure 2.1.

The various air classifier designs can be categorized into three groups, namely, vertical, horizontal, and inclined. Vertical air classifiers are by far the most common type employed for refuse processing. Accordingly, air classification modeling efforts focused on the vertical design.

Air classification has been applied at a variety of positions in the RDF processing line. High quality fuels have been produced by processing schemes which employ air classification near the beginning, in the middle, or near the end of the processing line. Thus, the positioning of the air classifier in RDF processing schemes appears to be flexible, provided that the air classifier design and operating conditions are matched to the air classifier infeed conditions.

The key parameters that influence the performance of an air classifier include the air flowrate, the solids (refuse) flowrate, the cross-

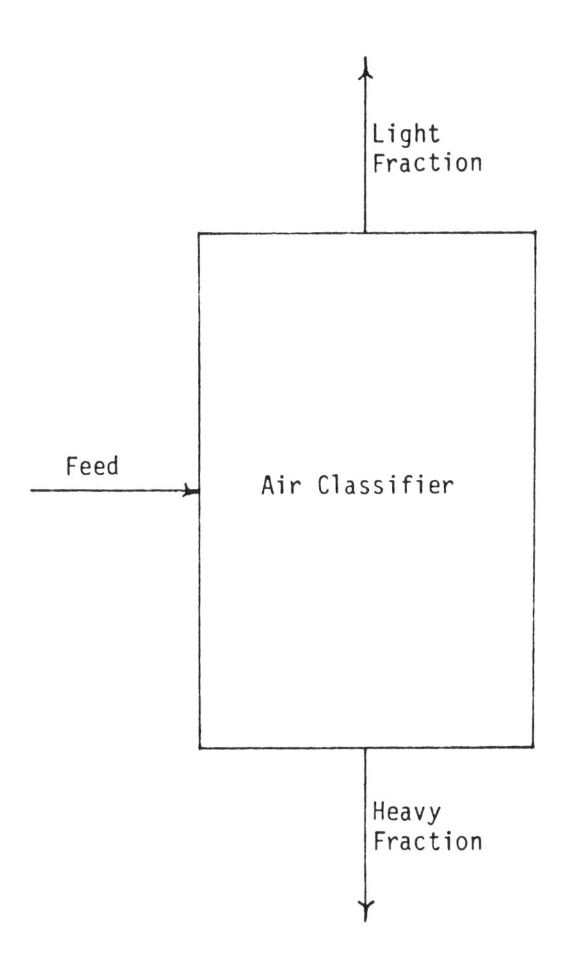

Figure 2.1. Block Diagram of Air Classification Process

sectional area of the air classifier, the column velocity, the feedstock composition, the feedstock moisture content, the size distribution of the feedstock, and the air density. Of these parameters, usually operational control can only be exerted over the air flowrate and the solids flowrate.

The air-to-solids ratio is defined as the ratio of the air and solids flowrates. This ratio has been demonstrated in numerous field tests to significantly affect the performance of an air classifier. There has been shown to exist a critical air-to-solids ratio above which the light fraction and heavy fraction splits are essentially constant and below which the percentage of material reporting to the light fraction drops off sharply. This point of transition is often referred to as the "choking" point.

Development of test methods for evaluating air classifier performance is only now taking place. One major reason for the lack of development of test methods is the vagueness in defining "light" and "heavy" materials. Some efforts have been made to devise performance parameters in terms of the recovery of paper and plastic and in terms of retained ash in the light fraction; however, these infrequently used parameters have little bearing on the system model developed for air classifiers in the present work. The model was tested in regard to: (1) its ability to predict the light fraction and heavy fraction splits; and (2) its ability to predict the light fraction and heavy fraction composition as compared to measured air classifier field test data.

2.2 DEVELOPMENT OF MASS BALANCE MODEL

2.2.1 Description of Model

The air classification model has been developed for a vertical air classifier. Based upon a given component size distribution of the feed to the air classifier, the component size distributions of the light fraction and of the heavy fraction are predicted. The governing relations that determine the predictions are functions of the following parameters: air flowrate, refuse flowrate, cross-sectional area of the air classifier, air density, material density, particle size, and particle shape. Accordingly, these parameters, along with the component size distribution of the feed, are the inputs to the model that was developed. A diagrammatic illustration of the air classifier model is shown in Figure 2.2.

Whether a particle that enters the air classifier will be carried by the flowing air stream to the light fraction or whether it will fall to the heavy fraction is determined by the forces exerted upon it. The two dominant forces acting upon the particle are the gravitational force and the drag force. The air velocity at which the drag force is equal to the gravitational force is referred to as the "terminal velocity." At this condition, the particle will neither settle nor rise, but rather will float in the column of air. It follows that if the air velocity is greater than the terminal velocity of the particle, the particle will be carried to the light fraction stream. Conversely, if the air velocity is less than the terminal velocity of the particle, then the particle will fall to the heavy fraction stream.

The core of the air classification model is the determination of the terminal velocity for each component and size element in the feed matrix and the subsequent comparison of that terminal velocity to the air velocity. The drag force on a given particle may be expressed as,

$$D = 0.5\, C_D\, A_{xp}\, \rho_a\, V_o^2 \tag{1}$$

where:

D = drag force;
C_D = drag coefficient;
ρ_a = air density;
A_{xp} = cross-sectional area of the particle; and
V_o = column velocity.

The column velocity is defined as

$$V_o = \frac{Q_a}{A_{xac}} \tag{2}$$

where:

Q_a = volumetric air flowrate; and
A_{xac} = cross-sectional area of the air classifier.

The gravitational force on the particle is given by

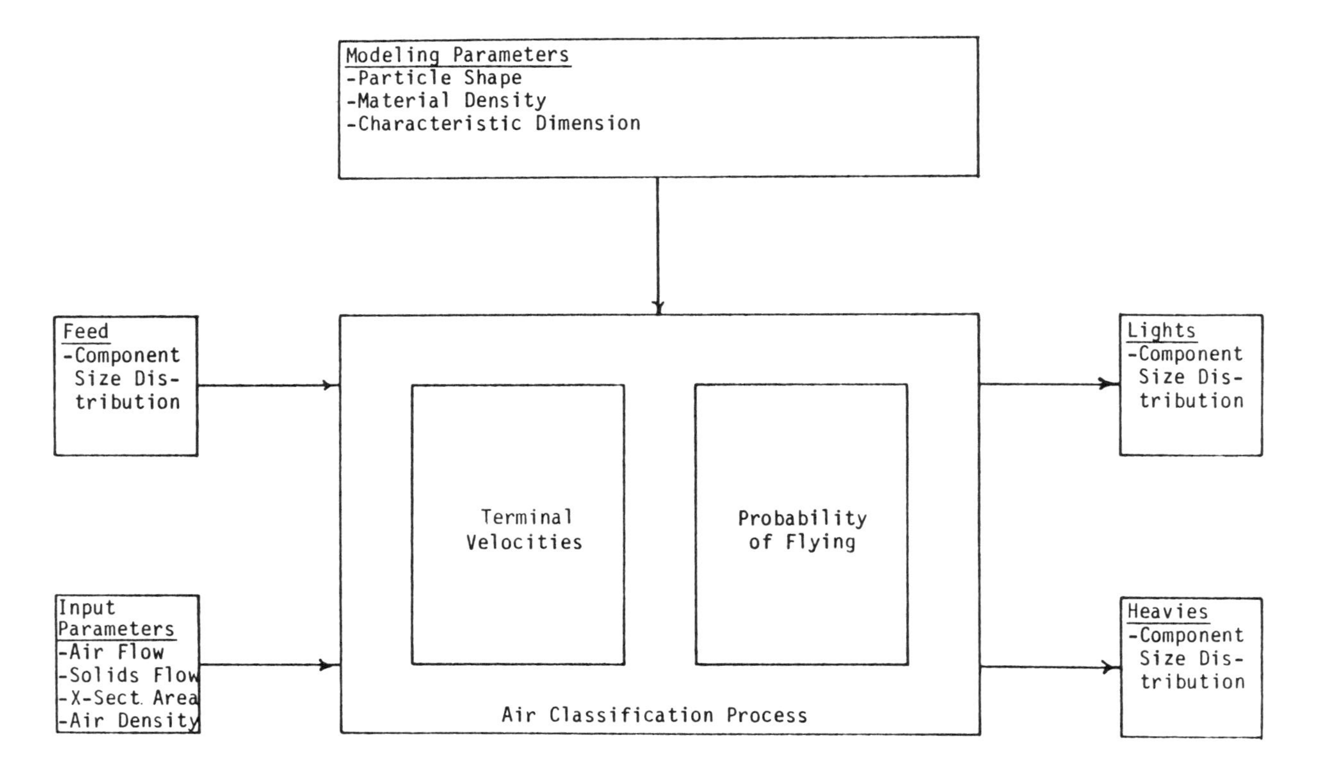

Figure 2.2. Block Diagram of Air Classification Model

$$W = \rho_s Yg \tag{3}$$

where:

W = gravitational force (i.e., weight);

ρ_s = particle density;

Y = particle volume; and

g = gravitational acceleration.

At equilibrium between the drag force and the gravitational force,

$$D = W \tag{4}$$

Substituting Eqs. 1 and 3 yields

$$0.5\ C_D\ A_{xp}\ \rho_a\ V_o^{\ 2} = \rho_s Yg \tag{5}$$

Solving for the terminal velocity, V_t,

$$V_t = \frac{2\ \rho_s Yg}{C_D A_{xp} \rho_a}^{0.5} \tag{6}$$

Thus, in order to calculate the terminal velocity for a particle, values must be determined for the particle density, the particle volume, the drag coefficient for the particle, the cross-sectional area of the particle, and the air density. The air density is a straightforward input. Similarly, the particle density can also be stated for a given component. The drag coefficient, particle volume, and particle cross-section are functions of the particle shape.

Four generic particle shapes were utilized in the air classification model: flakes, cylinders, splinters, and irregular-shaped particles. Irregular-shaped particles are generally modeled as cubes. The characteristics of the various particle shapes are summarized in Table 2.1. Modeling of the parameter values for these shapes is based upon information presented in Ref. 1. It has been shown that for the range of Reynolds numbers commonly encountered in air classifying shredded refuse, the drag coefficients for the particle shapes listed above are approximately constant. The estimated coefficients are 1.0 for plate-type

Table 2.1. Characteristics of Particle Shapes Utilized in Air Classification Modeling

Particle Designation	Aerodynamic Model	Dimensional Characteristics[a]	Drag Coefficient (C_D)	Cross-sectional Area (A_{xp})
Flake	Flat Plate	t << L	1.0	L^2
Cylinder	Cylinder	L < 15D	0.7	LD
Splinter	Cylinder	L > 15D	0.9	LD
Irregular	Cube	All sides of equal L	0.8	$2^{0.5}L^2$

[a] L = length; t = thickness; D = diameter.

particles, 0.7 for cylindrical particles, 0.9 for splinter-type particles, and 0.8 for cubical particles. Material properties and typical terminal velocities calculated for various waste components are presented in Table 2.2.

In order to account for the variations in particle characteristics (e.g., drag coefficients, densities, projected areas, and volumes) in a given component size category, a statistical formulation was employed in the model. It was assumed that the variations in particle characteristics in a given component size category follow a Gaussian distribution. The frequency function for the Gaussian distribution of the terminal velocity is given by

$$f(V_t) = \frac{1}{\sigma\ (2\pi)^{0.5}}\ e^{-(V_t - \overline{V}_t)^2/2\ \sigma^2} \tag{7}$$

where:

V_t = the terminal velocity of a given particle;
$\overline{V}_t$ = the average terminal velocity for the component size class; and
σ = the standard deviation.

Utilizing the standard normal variate, Z, given by

$$Z = \frac{V_t - \overline{V}_t}{\sigma} \tag{8}$$

Table 2.2. Material Properties and Typical Terminal Velocities for Various Waste Components

Waste Component	Moisture Content (percent)	Particle Density (kg/m^3)	Particle Designation	Typical Characteristic Dimension (t or L)(mm)	Terminal Velocity (m/s) as a function of t or L in mm	Typical Terminal Velocity (m/s)
Newsprint	10	560	Flake	8.9×10^{-2}	$3.0t^{0.5}$	0.9
	40	840	Flake	8.9×10^{-2}	$3.7t^{0.5}$	1.1
Ledger	10	758	Flake	1.0×10^{-1}	$3.5t^{0.5}$	1.1
	40	1,138	Flake	1.0×10^{-1}	$4.3t^{0.5}$	1.3
Corrugated S.W.[b]	10	192	Flake	3.7×10^{0}	$1.8t^{0.5}$	3.5
	40	320	Flake	3.7×10^{0}	$2.3t^{0.5}$	4.4
Corrugated L.B.[c]	10	650	Flake	3.2×10^{-1}	$3.2t^{0.5}$	1.8
	40	974	Flake	3.2×10^{-1}	$3.9t^{0.5}$	2.2
Polyethylene Film	3	912	Flake	5.8×10^{-1}	$5.8t^{0.5}$	4.4
Polyethylene	3	912	Irregular	5.8×10^{0} to 1.8×10^{1}	$3.6L^{0.5}$	8.7 to 15.3
Polyethylene Coated	10	746	Flake	7.4×10^{-1}	$3.5t^{0.5}$	3.0
	30	1,066	Flake	7.4×10^{-1}	$4.1t^{0.5}$	3.5
PVC Film	3	1,008	Flake	2.5×10^{-2}	$4.0t^{0.5}$	0.6
Lumber	12	480	Splinter	1.3×10^{1} to 2.0×10^{2}	$0.6L^{0.5}$	2.2 to 8.5
	30	603	Splinter	1.3×10^{1} to 2.0×10^{2}	$0.7L^{0.5}$	2.5 to 9.9
Plywood	12	552	Flake	3.8×10^{0}	$3.0t^{0.5}$	5.9

Table 2.2 (Cont'd)

Waste Component	Moisture Content (percent)	Particle Density (kg/m^3)	Particle Designation	Typical Characteristic Dimension (t or L)(mm)	Terminal Velocity (m/s) as a function of t or L in mm	Typical Terminal Velocity (m/s)
Textiles	5	242	Flake	1.3×10^0	$2.0t^{0.5}$	2.3
Rubber	3	1,773	Irregular	1.3×10^1	$5.0L^{0.5}$	18.0
	3	1,773	Flake	2.5×10^0 to 5.1×10^0	$5.3t^{0.5}$	8.4 to 12.0
Aluminum	0	2,688	Flake	1.3×10^{-1} to 4.1×10^{-1}	$6.6t^{0.5}$	2.4 to 4.6
	0	2,688	Irregular	2.5×10^0 to 5.1×10^1	$6.2L^{0.5}$	9.8 to 44.2
Aluminum Can	0	58	Cylinder	1.2×10^2	$0.6L^{0.5}$	6.6
Ferrous	0	7,840	Flake	1.3×10^{-1} to 2.8×10^{-1}	$11.2t^{0.5}$	4.0 to 5.9
	0	7,840	Irregular	2.5×10^0 to 5.1×10^1	$10.5L^{0.5}$	16.6 to 75.0
Ferrous Can	0	144	Cylinder	1.2×10^2	$0.9L^{0.5}$	9.9
Glass	0	2,400	Irregular	2.5×10^{-1} to 1.5×10^1	$5.8L^{0.5}$	2.9 to 22.5

[a] Ref. 1.
[b] S.W. = single wall (corrugated sandwich)
[c] L.B. = linerboard (1-ply corrugated)

Eq. 7 can be expressed as

$$f(V_t) = \frac{1}{(2\pi)^{0.5}} e^{-Z^2/2} \tag{9}$$

From the definition of the coefficient of variation, C_v, the Gaussian distribution may be reformulated without explicit use of the standard deviation. Eq. 8 becomes

$$Z = \frac{V_t - \bar{V}_t}{C_v \bar{V}_t} \tag{10}$$

Of interest is the total fraction of all particles in a given component size category whose terminal velocities are less than the air velocity. This fraction is given by the cumulative distribution function, ϕ, which can be expressed in equation form as

$$\phi\ (Z) = \frac{1}{(2\pi)^{0.5}} \int_{-\infty}^{Z} e^{-Z^2/2}\, dZ \tag{11}$$

with the standard normal variate given by

$$Z = \frac{V_o - \bar{V}_t}{C_v \bar{V}_t} \tag{12}$$

Thus, the mass fraction of material in a given component size category that reports to the light fraction is given by

$$mf_l(i,j) = \phi(Z(i,j))\ mf_f(i,j) \tag{13}$$

where:

- $mf_l(i,j)$ = the mass fraction of a given component size category in the light fraction;
- $mf_f(i,j)$ = the mass fraction of a given component size category in the feed;
- i = the component index; and
- j = the size index.

The mass fraction of material in a given component size category that reports to the heavy fraction is given by

$$mf_h(i,j) = mf_f(i,j) - mf_l(i,j) \tag{14}$$

where:

$mf_h(i,j)$ = the mass fraction of a given component size category in the heavy fraction.

Typically, Eq. 11 is solved by means of tables. Alternatively, it may be solved numerically.

2.2.2 Important Assumptions

The major assumptions made in the development of the air classification model are summarized in Table 2.3.

2.2.3 Accuracy of Model

A simulation of the model and its accuracy is presented in Figure 2.3. The model predictions are compared with field test data reported in Ref. 2. The columns of the matrices represent various size categories, and the rows represent various component categories. The four component categories shown are those used in the air classifier field test work.

Shown in the figure are the component size distributions of the feed, the calculated light fraction, the measured light fraction, the

Table 2.3. Air Classifier Model Assumptions

1. Vertical air classifier.
2. Air-to-solids ratio is above the "choking" point.
3. Component size categories can be assigned characteristic shapes, densities, particle cross-sections, and particle volumes.
4. Drag coefficients are constant (for a given particle shape) in the range of Reynolds Numbers encountered in typical air classifier operating conditions.
5. Variations in particle characteristics, such as drag coefficients, densities, projected areas, and volumes, follow a Gaussian distribution.
6. Particle-to-particle interactions are neglected.

Simulation Parameters

Air Flowrate = 2.568 m^3/sec
Mass Feedrate = 2.37 Mg/hr
A/C Cross-Sectional Area = .39 m^2
Column Velocity = 6.5846 m/sec
Air-to-Solids Ratio = 4.993
Column Loading = 6.0769 Mg/m^2/hr

	Size Class (mm)				
	-1.3	-9.5 +1.3	-16 +9.5	-25 +16	+25
Feed (Mass Fraction x 100)					
Paper/Plastic	0	9.448	11.129	11.44	9.173
Aluminum	0	.056	.442	.43	1.253
Ferrous	0	.617	.809	1.707	2.336
Other	14.185	31.891	2.766	1.232	1.086
Calculated Light Fraction (Mass Fraction x 100)					
Paper/Plastic	0	9.448	11.129	11.44	9.173
Aluminum	0	.0553	.4365	.4246	1.2373
Ferrous	0	.1447	.1667	.3338	.4187
Other	12.6025	13.047	.8544	.336	.2387
Measured Light Fraction (Mass Fraction x 100)					
Paper/Plastic	0	9.154	10.869	11.231	8.973
Aluminum	0	.043	.414	.345	.884
Ferrous	0	.51	.262	.656	0
Other	13.743	26.698	1.54	.853	.825
Error in Calculated Light Fraction (%)					
Paper/Plastic	0	3.2117	2.3921	1.8609	2.2289
Aluminum	0	28.6031	5.4275	23.0782	39.9686
Ferrous	0	-71.6334	-36.3756	-49.1189	0
Other	-8.299	-51.1311	-44.5191	-60.6071	-71.0631

Figure 2.3. Air Classification Model Simulation

	Size Class (mm)				
	-1.3	-9.5 +1.3	-16 +9.5	-25 +16	+25
Calculated Heavy Fraction (Mass Fraction x 100)					
Paper/Plastic	0	0	0	0	0
Aluminum	0	.0007	.0055	.0054	.0157
Ferrous	0	.4723	.6423	1.3732	1.9173
Other	1.5825	18.844	1.9116	.896	.8473
Measured Heavy Fraction (Mass Fraction x 100)					
Paper/Plastic	0	.294	.26	.209	.2
Aluminum	0	.013	.028	.085	.369
Ferrous	0	.107	.547	1.051	2.336
Other	.442	5.193	1.226	.379	.261
Error in Calculated Heavy Fraction (%)					
Paper/Plastic	0	-100	-100	-100	-100
Aluminum	0	-94.6103	-80.2493	-93.6705	-95.7514
Ferrous	0	341.4301	17.4231	30.6584	-17.9231
Other	258.0401	262.8728	55.9212	136.406	224.6247

Comparison of Predicted Performance

	Calculated	Measured	Error (%)
Lights Split (%)	71.4862	87	-17.832
Heavies Split (%)	28.5138	13	119.3371

Figure 2.3 (Con't)

calculated heavy fraction, and the measured heavy fraction. Also shown are the percent errors in each of the component size categories for the caculated light fraction and the calculated heavy fraction. Finally, a comparison of calculated and measured light fraction and heavy fraction splits are shown.

2.3 ENERGY REQUIREMENTS

The power to operate an air classifier (P_t) comprises two components, namely; blower power (P_b) and auxiliary power (P_a).

$$P_t = P_b + P_a \tag{15}$$

Auxiliary power, or the power used to drive auxiliary equipment varies from system to system because of variations in the system design. Auxiliary equipment may include air locks, internal conveyors, vibrators, auxiliary blowers, etc. Midwest Research Institute and Cal Recovery Systems reported on the operating characteristics of seven air classifiers in MSW processing operations [2] and presented data that supports the approximation,

$$P_a = 0.1\ P_t \tag{16}$$

It follows from Eqs. 15 and 16 that,

$$P_t = \frac{P_b}{0.9} \tag{17}$$

The electrical power required to drive a blower is,

$$P_b = \frac{(p_s + p_v)\ Q_a}{1000\ \eta} \tag{18}$$

where:

η = efficiency of blower and motor;
P_b = blower power (kW);
p_s = static pressure (Pa);
p_v = velocity pressure (Pa); and
Q_a = volumetric air flowrate (m^3/s).

The efficiency varies with load. Based on field measurements of energy consumption by the Baltimore County air classifier [2], the efficiency of the blower motors has been assumed to be 60 percent. The volumetric air flowrate (Q_a) is an input to the power model.

The velocity pressure is given by,

$$p_v = 1/2 \rho_a V_o^2 \tag{19}$$

where:

p_v = velocity pressure (Pa);
ρ_a = density of air (kg/m^3); and
V_o = column velocity of air (m/s).

Approximating the density of air (ρ_a) by 1.2 kg/m^3 yields,

$$p_v = 0.60 V_o^2 \tag{20}$$

The velocity is an input to the power model.

The static pressure depends on the dimensions of the air classifier and ductwork; air velocity; shape and roughness of the air classifier and ductwork; presence or absence of baghouses, air locks, cyclones, dampers, and other equipment in the system; presence of solids in the air classifier; and other factors that may obstruct or resist the flow of air. Data in Ref. 2 show that a pressure drop across a blower of about 2.6 kp (250 mm water gauge) is typical. However, it is common to have air pass through two blowers in series. One blower supplies air to the air classifier, while the second blower is used to force air through a baghouse or to recirculate it. Based upon the data in Ref. 2 and engineering judgment, the total static pressure drop for the energy model has been taken to be 5.2 kp (500 mm W.G. or 20 in. W.G.).

Using the approximations and assumptions given above, Eq. 18 can be written as,

$$P_b = \frac{(5200 + 0.60 V_o^2) Q_a}{600} = (8.67 + .001 V_o^2) Q_a \tag{21}$$

Eq. 17 then yields,

$$P_t = (9.63 + 0.001\ V_o^2)\ Q_a \qquad (22)$$

where the units of the quantities are defined as: (1) P_t in kW; (2) V_o in m/s; and (3) Q_a in m^3/s.

The specific energy E_s is,

$$E_s = \frac{P_t}{\dot{m}} \qquad (23)$$

where the units of the quantities are defined as: (1) E_s in kWh/Mg; (2) P_t in kW; and (3) $\dot{m}$ in Mg/hr.

If the ratio (R) of the mass of air to the mass of solids is known and the density of air (ρ_a) = 1.2 kg/m^3, then the air flowrate (Q_a) can be expressed in terms of the flowrate of solids ($\dot{m}$),

$$Q_a = 0.23\ R\ \dot{m} \qquad (24)$$

where the units of the quantities are defined as: (1) Q_a in m^3/s; and (2) $\dot{m}$ in Mg/hr.

The power consumption of three air classification systems tested by Midwest Research Institute [2] is compared to the power predicted by Eq. 22 in Table 2.4.

2.4 REFERENCES

1. Diaz, L.F., Savage, G.M., and Golueke, C.G., Resource Recovery from Municipal Solid Wastes, CRS Press, Inc., 1982.

2. Hopkins, V., Simister, B.W. and Savage, G., Comparative Study of Air Classifiers, Final Report, EPA Contract No. 68-03-2730, 1980.

Table 2.4. Comparison of Measured Power to Predicted Power for Three Commercial Air Classifiers

Air Classifier Location	Volumetric Air Flow Q_a (m^3/s)	Column Velocity V_o (m/s)	Measured Power (kW)	Predicted Power[a] (kW)
Baltimore, Maryland	27	31	315	286
	31	35	325	337
	37	43	356	425
Akron, Ohio	26	22	284	263
	29	25	283	297
	30	26	283	309
Ames, Iowa	12	19	146	120
	13	20	155	130
	13	21	158	131

[a]Predictions are calculated using Eq. 22 in the text.

3. Trommel Screening Model

3.1 BACKGROUND

Screening is used in RDF preparation for removing inorganic materials from the fuel fraction and in some cases for particle size control. The use of screening in an RDF production line generally correlates well with fuel quality. Improvement of fuel quality can be accomplished because the particle sizes of the combustible materials (e.g., paper and plastic) tend to be relatively large in comparison to the particle sizes of the inorganic materials. Removal of inorganic materials decreases the ash and moisture contents of the RDF and thereby increases the heating value of the material.

Three different types of screens have been used for RDF processing: trommel screens, disc screens, and flatbed screens. The focus of this section is on the trommel screening process. At the present time, far more work has been performed in the analysis of trommel screening for refuse processing applications than has been performed in the analysis of any other screening process. However, many of the principles discussed in the trommel screening model are also applicable to disc screening and flatbed screening.

Trommel screens are rotating cylindrical screens set at an inclination to the horizontal. Feed material is introduced at the upper end, and is conveyed down the length of the screen by means of the tumbling action imparted on the material. Undersize material passes through the apertures in the screen and is transported by a conveyor belt. Oversize material and any undersize material that do not pass through the screen exit at the lower end of the screen. This material is then transported by a conveyor belt. A simplified block diagram of a trommel screening process is shown in Figure 3.1.

The material that passes through the screen is commonly referred to as either the "undersize" or the "unders." The material that does not pass through the screen is commonly referred to as the "oversize" or the "overs." However, a distinction must be made between the true oversize

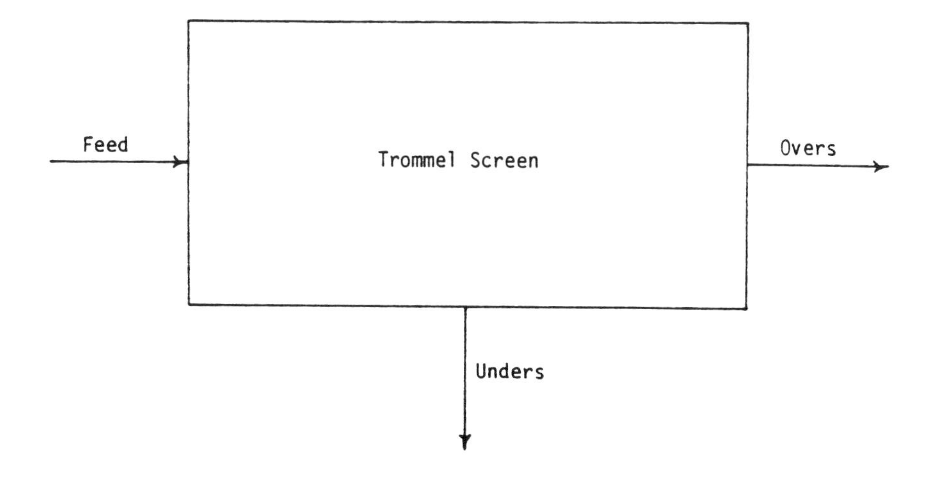

Figure 3.1. Block Diagram of Trommel Screening Process

and the material that does not pass through the screen, inasmuch as some undersize material is present in the material that exits the screen at the downstream end. For the purposes of the screening discussion, the terms "unders" and "overs" are used to describe the split streams leaving the screen.

Screening can be employed in a variety of places in the processing sequence in order to accomplish a set goal, if the screen has been properly designed for the particular operating conditions. For example, trommel screens can be employed as the first stage in a refuse processing line. In such a position, the screens are commonly referred to as "pre-trommels," implying screening prior to size reduction of the material. Thus, the material being screened is raw MSW.

Most applications of trommels in the resource recovery industry have been for processing shredded MSW. Such trommels are sometimes referred to as post-trommels. In some plants, the trommel is located following air classification; while in other plants, the trommel is located prior to air classification.

The key parameters affecting the performance of a trommel screening operation can be categorized as construction parameters, operating parameters, and feed characteristics. Typically the only operating parameters over which control can be exerted are the feed rate and the rotational speed of the screen. Feed rate is the dominating parameter governing screening performance.

The parameter most commonly used for the characterization of screening performance is the screening efficiency. This parameter represents the percentage of undersize material entering the screen that passes through the apertures in the screen to the undersize fraction. In equation form, screening efficiency (η) can be formulated in several ways. The following form is employed in the development of the trommel screening model:

$$\eta = \frac{\dot{m}_u}{\dot{m}_u + U_o \dot{m}_o} \qquad (1)$$

where:

$\dot{m}_u$ = flowrate of the undersize fraction;
$\dot{m}_o$ = flowrate of the oversize fraction; and
U_o = fraction of true undersize material in the oversize fraction.

3.2 DEVELOPMENT OF MASS BALANCE MODEL

3.2.1 Description of Model

Although the trommel screening process appears to follow a cyclical pattern of tumbling material, the behavior of individual particles in the process involves highly irregular motion from one cycle to the next. Modeling of the forces acting on the particles becomes complex if the depth of the bed of material on the screen is taken into account. Efforts at modeling the particle dynamics at different locations in the screen were presented by Glaub, et al [1]. These models involve the generation of a system of simultaneous differential equations which are solved in sequential time steps as the material traverses the length of the screen.

In the current work effort, the objective is to develop models that lend themselves toward computerized integration. Thus, a trommel screening model of intermediate complexity is developed, i.e., more complex than a transfer function model but less complex than series of simultaneous differential equations.

The model is structured such that a component size feed matrix is input to the model. The model subsequently predicts the component size matrices of the overs stream and of the unders stream. The other inputs to the model are the screen construction parameters and the screen operating parameters. The screen construction parameter inputs are the diameter, the effective screen length (i.e., the perforated length), the inclination angle, the aperture size, and the open area fraction. The operating parameter inputs are the feed rate and the departure angle. For purposes of reducing the complexity of the model, the departure angle is specified as an input parameter rather than calculated. The departure angle is the angle measured from vertical to the point at which the material detaches from the screen. The departure angle is a function of the rotational speed of the screen (an operating parameter) and of the presence and configuration of lifters inside the screen. In addition to the component size matrices of the overs and unders streams, outputs from the model include the screening efficiency and the overs and unders splits. A diagrammatic illustration of the model is shown in Figure 3.2.

There are two primary aspects of particle behavior in a screening process that determine the screening performance and the characteristics of the output streams from the screen: (1) the particle dynamics and (2) the probability of passage for a given particle. The particle dynamics determine the number of contacts between a particle and the screen surface. The probability of passage concerns the mechanism by which particles pass or do not pass through the apertures in the screen.

Of the several relations that have been proposed for predicting the number of contacts between a particle and the screen [1], the following relation is used in the current model:

$$N_c = \frac{L}{4D \operatorname{Tan} \beta \operatorname{Cos} \alpha \operatorname{Sin}^2 \alpha} \tag{2}$$

where:

- N_c = calculated number of contacts;
- L = effective screen length;
- D = diameter;
- β = inclination angle; and
- α = departure angle.

To account for reduced contact with the screen surface as the screen loading increases (e.g., material falling unto other material that is already covering the screen surface), Eq. 2 is modified as follows:

$$N_e = aN_c^{\,b} \tag{3}$$

where:

- N_e = the effective number of contacts with the bare screen surface;
- a = a modeling parameter that may be a function of feed rate or holdup; and
- b = a modeling parameter that accounts for multiple contacts per drop.

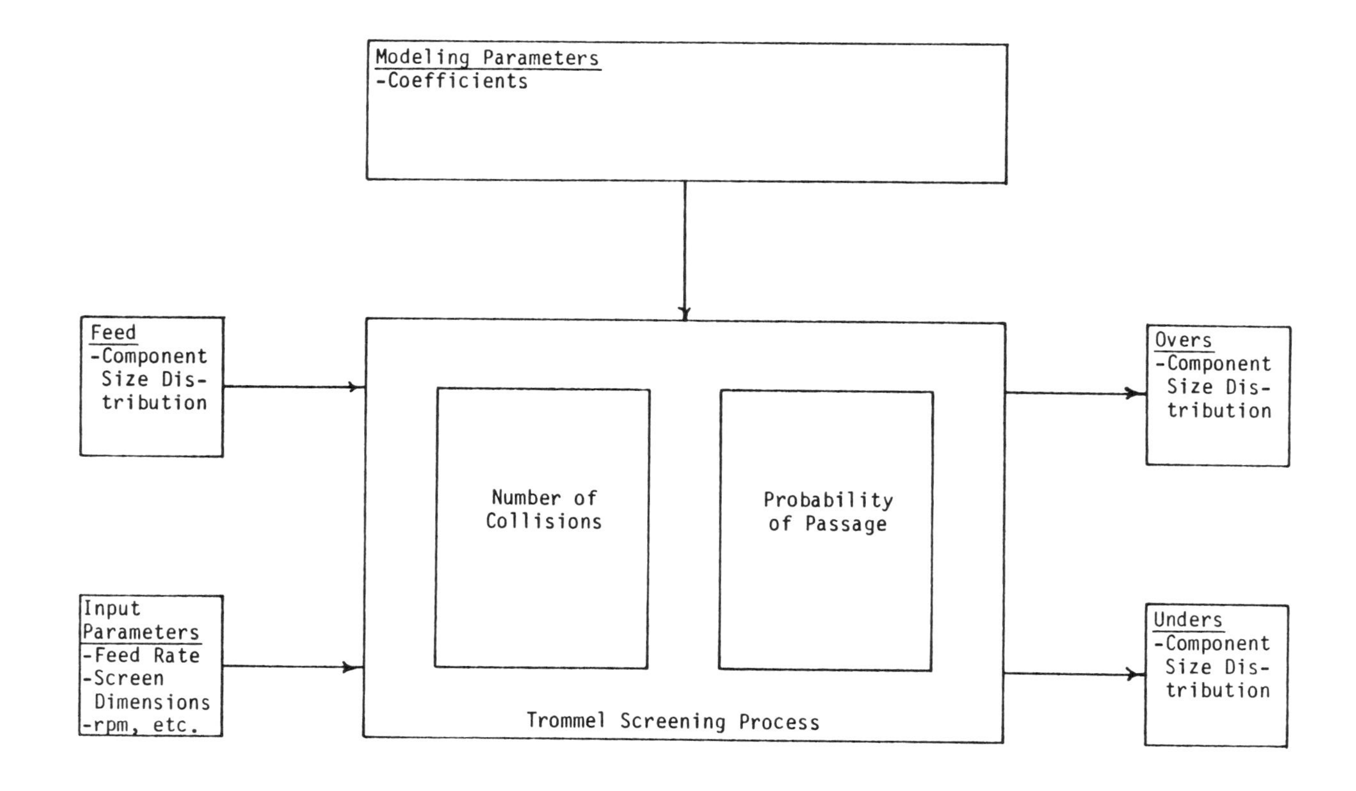

Figure 3.2. Block Diagram of Trommel Screening Model

Both a and b are m x n matrices for an m component and n size class system. It follows that N_e is also an m x n matrix. Thus, the model accounts for a variation in number of contacts as a function of component and size of class.

The probability of passage is based upon the simple probability of passage:

$$P_o = \left(1 - \frac{D_p}{D_a}\right)^2 f_a \quad (4)$$

where:

P_o = the probability of passage;
D_p = the particle size;
D_a = the aperture size; and
f_a = the open area fraction.

The current model also incorporates an additional modeling parameter coefficient in the probability of passage relation such that

$$P = cP_o \quad (5)$$

where:

P = the modified probability of passage; and
c = a modeling parameter.

In the model, both P and c are m x n matrices. Therefore, each element in the m x n feed matrix has its own probability of passage associated with it.

The mass fraction of material in a given component size category that reports to the oversize fraction is given by

$$mf_v(i,j) = (1-P)^{N_e}\, mf_f(i,j) \quad (6)$$

where:

$mf_v(i,j)$ = the mass fraction of a given component size category in the overs stream;
$mf_f(i,j)$ = the mass fraction of a given component size category in the feed;

i = the component index; and

j = the size index.

The mass fraction of material in a given component size category that reports to the undersize fraction is given by

$$mf_u(i,j) = (1-(1-P)^{N_e})\, mf_f(i,j) \quad (7)$$

where:

$mf_u(i,j)$ = the mass fraction of a given component size category in the unders stream.

3.2.2 Important Assumptions

The major assumptions in the development of the trommel screening model are summarized in Table 3.1.

3.2.3 Accuracy of Model

A simulation of the model and its accuracy is presented in Figure 3.3. The model predictions are compared with field test data reported by Hennon, *et al* [2]. The columns of the matrices represent various size categories, and the rows represent various component categories. The

Table 3.1. Trommel Screening Model Assumptions

1. The trommel screen is operated in centrifugal action, i.e., the material cycles through stages of lifting and falling. The model does not account for operation of a trommel screen in kiln action.
2. Drag forces on the particle in the falling phase are negligible.
3. Slippage and tumbling of material on the screen are neglected.
4. No size reduction occurs in the screen. Therefore, the sum of the mass of material in a given component size category in the overs and unders streams is equal to the mass of material in that same component size category in the feed.
5. A falling particle recontacts the screen at right angles to the screen surface.
6. The material is well-mixed.

Simulation Data

Feedrate = 2.64 Mg/hr
Rotational Speed = 6 rpm
Effective Screen Length = 7.4 m
Diameter = 3.7 m
Aperture Size = 25.4 mm
Open Area Fraction = .465
Inclination Angle = 3 degrees
Departure Angle = 45 degrees

	Size Class (mm)				
	-1.3	-5.1 +1.3	-12.7 +5.1	-25.4 +12.7	+25.4
Feed (Mass Fraction x 100)					
Paper/Plastic	5.41	13.59	21.99	10.3	15.9
Other Organics	5.42	8.25	1.7	.78	.85
Glass	5.41	3.89	.04	0	0
Other Inorganics	5.42	.34	.1	.08	.52
Calculated Oversize Fraction (Mass Fraction x 100)					
Paper/Plastic	0	.0002	.0751	4.7842	15.9
Other Organics	0	.0001	.0058	.3623	.85
Glass	0	0	.0001	0	0
Other Inorganics	0	0	.0003	.0372	.52
Measured Oversize Fraction (Mass Fraction x 100)					
Paper/Plastic	.03	.04	.49	3.02	15.9
Other Organics	.03	.08	.06	.05	.85
Glass	.03	0	0	0	0
Other Inorganics	.03	0	0	0	.52
Error in Calculated Oversize Fraction (%)					
Paper/Plastic	-99.9951	-99.6233	-84.6704	58.4176	0
Other Organics	-99.9951	-99.8857	-90.3217	624.599	0
Glass	-99.9951	0	0	0	0
Other Inorganics	-99.9951	0	0	0	0

Figure 3.3. Trommel Screening Model Simulation

	Size Class (mm)				
	-1.3	-5.1 +1.3	-12.7 +5.1	-25.4 +12.7	+25.4
Calculated Undersize Fraction (Mass Fraction x 100)					
Paper/Plastic	5.41	13.5898	21.9149	5.5158	0
Other Organics	5.42	8.2499	1.6942	.4177	0
Glass	5.41	3.89	.0399	0	0
Other Inorganics	5.42	.34	.0997	.0428	0
Measured Undersize Fraction (Mass Fraction x 100)					
Paper/Plastic	5.38	13.55	21.5	7.28	0
Other Organics	5.39	8.17	1.64	.73	0
Glass	5.38	3.89	.04	0	0
Other Inorganics	5.39	.34	.1	.08	0
Error in Calculated Undersize Fraction (%)					
Paper/Plastic	.5576	.2941	1.9297	-24.2337	0
Other Organics	.5566	.9781	3.3045	-42.7808	0
Glass	.5576	-.0011	-.3416	0	0
Other Inorganics	.5566	-.0011	-.3416	-46.4487	0

Comparison of Predicted Performance

	Calculated	Measured	Error (%)
Screening Efficiency (%)	93.6347	95.6	-2.0557
Oversize Split (%)	22.5454	21.13	6.6984
Undersize Split (%)	77.4546	78.87	-6.6984

Figure 3.3 (Con't)

four component categories shown are those used in the trommel screening field test work.

Shown in the figure are the component size distributions of the feed, the calculated overs stream, the measured overs stream, the calculated unders stream, and the measured unders stream. Also shown are the percent errors in each of the component size categories for the calculated overs stream and the calculated unders stream. Finally, a comparison of calculated and measured screening efficiencies and overs and unders splits are shown.

3.3 ENERGY REQUIREMENTS

The energy requirements for trommel screening have been developed based upon actual test data. Early efforts to develop a theoretical governing equation for trommel energy requirements were terminated when the analysis showed that the theoretical values were an order of magnitude less than the values measured during the Baltimore Co. trommel study [2].

CRS used the test data from the Baltimore County trommel study to derive the energy requirements for trommel screening. Using the available data the power consumption of the trommel screen has been found to conform to an equation of the form:

$$P = 3.5 + 0.05\,\dot{m} \qquad (8)$$

where:

P = power (kW); and

$\dot{m}$ = mass throughput rate (Mg/hr).

Based upon the test data for the Baltimore trommel screen, the rated throughput of the unit is estimated to be 10 Mg/hr. From Eq. 10 the corresponding power consumption is 4 kW. The specific energy consumption is therefore 0.4 kWh/Mg.

Lacking further published data on the power requirements of trommel screens used in MSW processing plants, the value of 0.4 kWh/Mg is used as the energy requirement for the trommel screens handling refuse. The value of 0.4 kWh/Mg is within the range of values CRS has calculated from cursory measurements of energy usage of pilot-scale trommel screens. The value appears reasonable for both pre- and post-trommels.

3.4 REFERENCES

1. Glaub, J.C., D.B. Jones, J.U. Tleimat, and G.M. Savage, Trommel Screen Research and Development for Applications in Resource Recovery, Final Report under U.S. DOE Contract No. DE-AC03-79CS20490, Oct. 1981.

2. Hennon, G.J., D.E. Fiscus, J.C. Glaub, and G.M. Savage, An Economic and Engineering Analysis of a Selected Full-Scale Trommel Screen Operation, Final Report under U.S. DOE Contract No. DE-AC03-80CS24330, Oct. 1983.

4. Ferrous Separation Model

4.1 BACKGROUND

The recovery of ferromagnetic waste materials through the use of magnets has been practiced extensively in automobile salvage operations. The practice was extended to municipal solid waste initially when refuse was shredded prior to landfilling and later when refuse was processed for material or energy recovery. Magnetic separators have been used in refuse composting plants and incineration plants for salvage and, as is the case in composting, for improving the quality of the finished product.

The two basic types of magnetic separators for recovering ferromagnetic materials from municipal solid waste are the overhead belt magnet and the drum magnet. These pieces of equipment are manufactured in the form of electromagnets or permanent magnets. For the purpose of the work performed herein, the overhead belt was selected for modeling because of the fact that most of the data available in the literature were obtained with overhead belts.

Magnetic separators can be, and have been, installed at a number of locations within material recovery facilities. During the first stages of development of material and energy recovery from solid waste it was generally believed that size reduction prior to magnetic separation would be a sufficient step for producing a clean, acceptable product. Unfortunately, the expectations were much too high because, in most cases, impurities present in the recovered ferrous and the bulk density of the recovered product rendered the material unacceptable.

Experience accumulated thus far indicates that other unit processes must be included in the overall design in order to produce a salable ferrous scrap. Depending upon the type of end-user, size distribution especially that of the light gauge ferrous materials, is a critical factor as well.

The most important parameters that determine the performance of a magnetic separator are: (1) flux density and the rate of change of flux density (gradient) at the object, (2) size and shape of material to be recovered, and (3) distance between magnetic separator and material. Other parameters that affect the performance of a magnetic separator but that are interrelated include bulk density, depth of burden, and speed of conveyor belt carrying feedstock to the magnetic separator. In general, using a particular type of magnetic separator and a fixed speed for the conveyor belt carrying the feedstock, the quality or purity of the recovered magnetic metals increases as the depth of the burden decreases. A reduction in the depth of burden decreases the chances for any impurities to become attached to the metal.

A schematic diagram of a magnetic separator is presented in Figure 4.1. The figure shows the inputs to and outputs from the unit process. In this particular case, the feed to the magnetic separator is the shredded, air classified, heavy fraction of MSW. Mass flowrate, composition, and bulk density of the feed would have to be known. In addition, the size distribution of the magnetic metals would have to be prescribed.

The magnetic separator operates as a binary device thus the feedstock is separated into two streams: (1) magnetics (Fe); and (2) non-magnetics (Non-Fe). The quality of the recovered material can be expressed as the ratio of the mass of magnetic metals in the accepts (Fe_2) to the total mass of the accepts. The recovery of magnetic metals in the accepts is defined as the ratio of the mass of magnetic metals in the accepts to the mass of magnetic metals in the feed [1].

The model presented herein is based on an approach commonly followed by designers and manufacturers of magnetic separators. This approach is an attempt to simplify the complexities involved in the theory of magnetism.

4.2 DEVELOPMENT OF MASS BALANCE MODEL

The force per unit volume in the z-direction (F_z) developed between two magnetic surfaces separated by a gap can be described as follows:

$$F_z = -\rho_m \frac{\partial A}{\partial z} - \frac{H^2}{8\pi} \frac{\partial \mu}{\partial z} + \frac{\partial}{\partial z} \left(\frac{H^2}{8\pi} \rho \frac{\partial \mu}{\partial \rho} \right) \qquad (1)$$

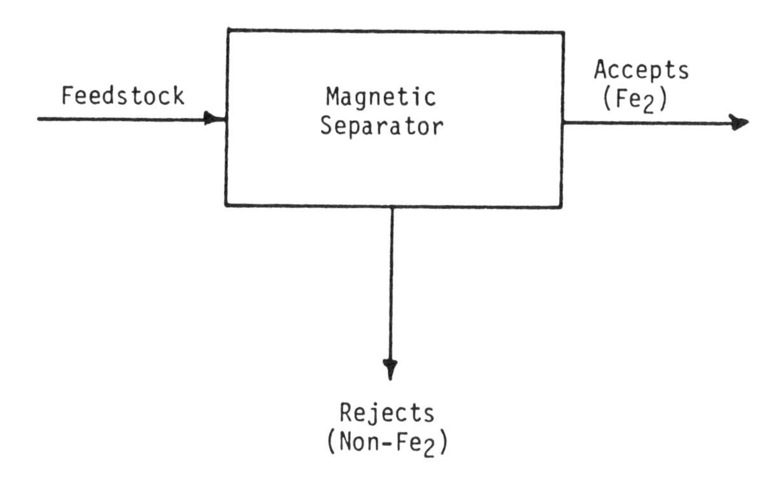

Figure 4.1. Schematic Diagram of Magnetic Separation

where:

ρ_m = density of magnetic poles;
A = magnetic potential;
H = magnetic field intensity;
μ = permeability; and
ρ = mass density.

Assuming that

$$\mu - 1 = C\rho \tag{2}$$

(where C is a constant) and that the permeability of the material does not change with distance (i.e., $\partial\mu/\partial z = 0$), Eq. 1 can be written as follows:

$$F_z = -\rho_m \frac{\partial A}{\partial z} + \frac{(\mu-1)}{4\pi} H \frac{\partial H}{\partial z} \tag{3}$$

The major portion of magnetic metals present in the waste stream do not have permanent magnetism. Therefore, the mathematical expression describing the magnetic force per unit volume acting on the particle can be further simplified as follows:

$$F_z = \frac{(\mu-1)}{4\pi} H \frac{\partial H}{\partial z} \tag{4}$$

The major forces acting on a magnetic particle resting on a conveyor belt moving past a magnetic separtor are:

$$F_T = F_m - F_g - F_b \tag{5}$$

where:

F_T = resultant force;
F_m = magnetic force;
F_g = gravitational force; and
F_b = force due to weight of burden.

so the total (resultant) force can be written as follows:

$$F_T = V_p \rho_p \frac{\partial^2 z}{\partial t^2} \tag{6}$$

$$= V_p \frac{\mu-1}{4\pi} H \frac{\partial H}{\partial z} - V_p \rho_p g - V_b \rho_b g \tag{7}$$

$$= V_p \rho_p \frac{\partial^2 z}{\partial t^2} \quad (8)$$

where:

V_p = volume of particle;
ρ_p = density of particle;
V_b = volume of burden on particle; and
ρ_b = density of burden.

The recovery of magnetic metals through magnetic separation depends on the differences in pathways followed by materials of different permeabilities as they pass through a magnetic field. Under these conditions, the materials are acted upon by magnetic, gravitational, and other forces. A full description of the pathway to be followed by a particle requires knowledge of a number of variables such as shape, size, density, and magnetic properties of the particle, as well as the magnitude of the forces acting on the particle as a function of location and time. A complete definition and solution of this type of problem is extremely complex. However, since $B = H/\mu$, the magnetic force described in Eq. 4 can be described as:

$$F_z = K\ B\ \frac{\Delta B}{\Delta z} \quad (9)$$

where:

K = constant;
B = flux density; and
$\Delta B/\Delta z$ = flux density gradient.

The term $B\ \Delta B/\Delta z$ is known as the Force Index (FI). The Force Index basically is a numerical expression of a magnet's ability to attract a magnetic object. The static force indices required to attract several types and shapes of magnetic metals have been determined. Similarly, the incremental Force Index needed to lift magnetic metals for various depths of burden has been measured.

In making a selection, the total Force Index (static plus incremental) required to lift the magnetic metal is matched to that developed by the magnetic separator at a certain distance above the conveyor belt.

The unit process described in this model consists of a magnetic belt installed above the pulley of a conveyor belt. Integral to the magnetic separation system are: (1) a feed conveyor from which the magnetic metals are removed, and (2) a conveyor belt to carry the recovered metals away from the magnetic separator.

In order to determine the quantity of magnetic metals that could be recovered from a given stream, the following steps would be taken:

1. Calculate the volumetric flowrate of the material in m^3/s.
2. Calculate the depth of burden (D_e) using the following formula:

$$D_e = (1.2)\left(\frac{C}{(W)(V)}\right) \tag{10}$$

where:

C = volumetric flowrate (m^3/s);
W = width of conveyor belt (m);
V = velocity of conveyor belt (m/s); and
D_e = depth of burden (m).

3. Calculate suspension height in meters.

$$S = D_e + 0.1 \tag{11}$$

4. Determine the smallest particle size to be recovered (for most practical purposes, 12 mm would be the minimum).
5. Read the minimum Force Index (as a function of burden) required to lift the smallest particle from Table 4.1. Assume the particle takes the shape of a cube.

Table 4.1. Minimum Force Index Required to Remove a Cube Using a Magnetic Belt Over Pulley Force Index (FI x 10^3)

Object Size (mm)	Burden Depth (mm)								
	50	100	150	200	250	300	350	400	450
13	93	99	105	111	117	123	129	136	141
19	83	87	91	96	100	104	109	114	119
25	78	81	85	88	92	97	99	105	107
50	70	72	74	76	78	80	82	85	87

6. Using the plot of FI vs. distance for a particular separator, determine if the unit will develop a sufficiently high FI to lift the particle at the suspension height S.

Using this approach implies that a sufficiently strong magnetic separator could be selected to remove all magnetic particles larger than the minimum stated. This situation, however, is not often realized particularly in the refuse processing industry. Contaminants inside of or attached to the magnetic metals change the recovery rates.

In order to account for variations in particle characteristics and, accordingly, in the force required to lift a particle in a given ferrous size category, a Gaussian distribution of the Force Index was employed. A description of the mathematics of the Gaussian distribution was presented in the discussion of the development of the mass balance model for air classification. In applying the Gaussian model, the first step is to determine the standard normal variate, Z, given by

$$Z = \frac{FI_m - \overline{FI}}{C_v \overline{FI}} \tag{12}$$

where:

- FI_m = Force Index developed by the magnet;
- $\overline{FI}$ = Average Force Index required to lift a particle in a given ferrous size category; and
- C_v = Coefficient of Variation of the Force Index required to lift a particle in a given ferrous size category.

The average Force Index, $\overline{FI}$, is obtained through use of Table 4.1 and Figure 4.2. Comparisons of the model to typical field data indicates that a value of 0.25 for the Coefficient of Variation yields reasonable efficiencies of separation, if more detailed data is unavailable for the material under consideration.

The total fraction of all particles in a given category whose force indices are less than the force index of the magnet (i.e., those particles which will be pulled to the magnet) is given by the cumulative distribution function, $\phi(Z)$. The cumulative distribution function is tabulated as a function of the standard normal variate in numerous statistics books and engineering mathematics books.

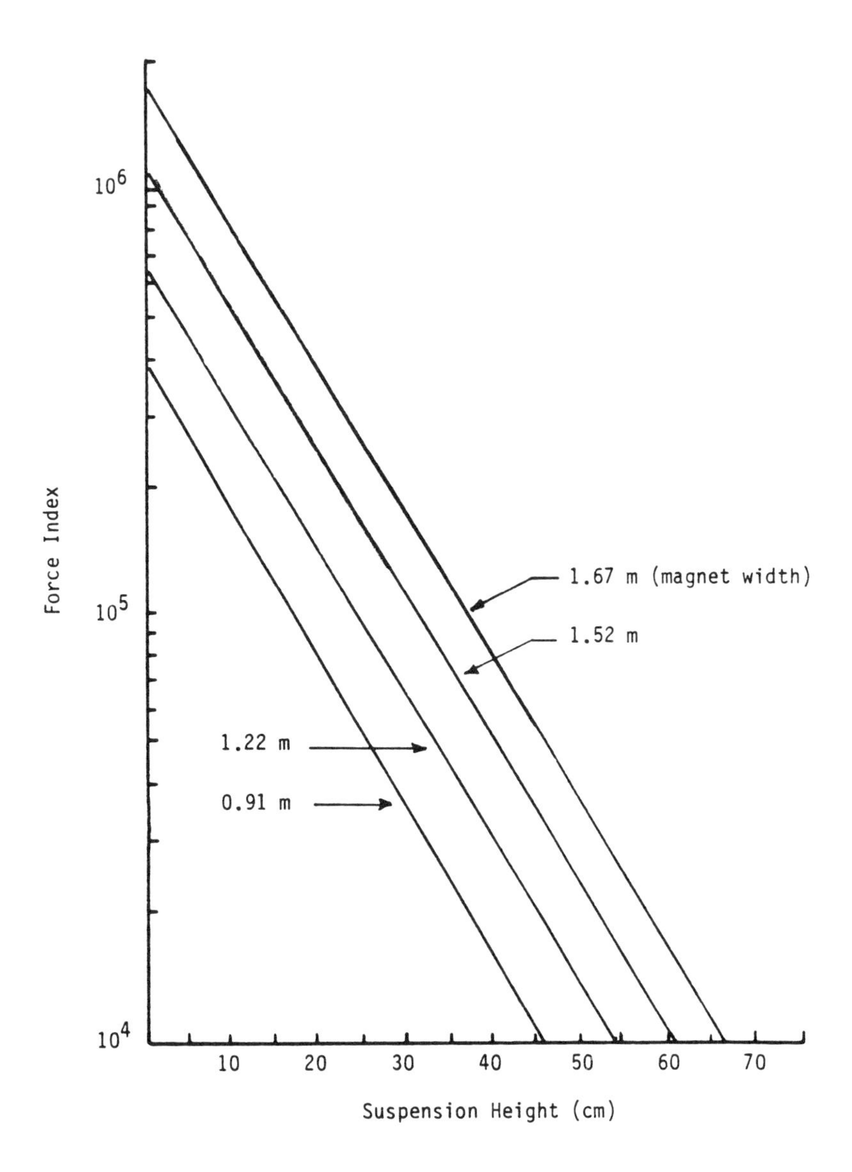

Figure 4.2. Force Index vs. Suspension Height for Several Commercial Magnetic Separators

The mass fraction of material in a given ferrous size category that reports to the recovered ferrous stream is given by

$$mf_{fe}(i) = \phi(Z(i))\, mf_f(i) \qquad (13)$$

where:

$mf_{fe}(i)$ = the mass fraction of a given ferrous size category that reports to the recovered ferrous stream;

$mf_f(i)$ = the mass fraction of a given ferrous size category in the feed; and

i = the size category index.

4.2.1 Example

A sample calculation is presented for a 0.91 m magnetic belt and an 8 cm depth of burden on a conveyor. From Figure 4.2, it is found that a 0.91 m magnetic belt develops a Force Index (FI_m) of approximately 95,000 at a suspension height of 18 cm. The suspension height is obtained from Eq. 11.

The size distribution of ferrous in a hypothetical feed is shown in Table 4.2 along with the average force indices ($\overline{FI}$) required to lift particles in the various categories. The average force indices were obtained from Table 4.1. The final column in Table 4.2 shows the calculated mass fraction of a given ferrous size category that will report to the recovered ferrous stream, as given by Eq. 13. The efficiency of the magnetic separation process in the example is 82.6 percent.

4.3 ENERGY REQUIREMENTS

The amount of energy consumed by a magnetic separator depends upon two factors: (1) type of magnet (i.e., electromagnet vs. permanent magnet), and (2) size of motor driving the belt.

In the range of 40 to 100 Mg of shredded waste per hour the specific energy required by an electromagnet is on the order of 0.26 kWh/Mg of waste. Similarly, the energy consumed by the motor driving the belt is about 0.1 kWh/Mg.

Therefore, a mathematical expression describing the energy requirements of a magnetic separation unit can be written as follows:

$$E = 0.36\,\dot{m} \tag{14}$$

where:

$\dot{m}$ = mass flowrate under magnetic belt.

Table 4.2. Sample Calculations for Magnetic Separation Model

Size Class (mm)	Typical Size (mm)	Ferrous in Feed (Weight Percent)	$\overline{FI}$	Recovered Ferrous (Weight Percent)
>38	50	0.630	71,000	0.574
22 - 38	25	0.220	79,500	0.172
16 - 22	19	0.090	85,000	0.061
10 - 16	13	0.040	96,000	0.019
<10	5	0.020	--	0.000
		1.000		0.826

4.4 REFERENCES

1. Alter, H., Testing and Evaluation Manual for Resource Recovery Plants, Prepared for the U.S. EPA under Contract 68-01-4423, Oct. 1978.

5. Densification Model

5.1 BACKGROUND

Densification renders RDF more suitable for storage, transportation, and in some cases, combustion by increasing the bulk density to about 500 kg/m^3, by binding easily lofted particles into a dense mass, and by producing fairly uniformly sized and uniformly shaped particles. There are several types of densification machines, but the type that has generally been selected for densifying RDF in commercial operations is a rotary-die extrusion mill - henceforth referred to as a "pellet mill." The product is cylindrical pellets generally having diameters of 12 to 25 mm and lengths of about 16 to 50 mm.

Because of the small particle size required for the proper operation of the pellet mills, they must be preceded in the processing line by size reduction. Generally, two stages of size reduction are required to achieve the requisite particle size for pelletizing. To form pellets with diameters of 12.5 mm and 25.4 mm, the feedstock should comprise particles with sizes not exceeding 19 mm and 38 mm, respectively. Air classification and/or screening often precede densification, not only to improve the quality of the RDF, but to decrease wear on the pellet mill. Metal recovery operations may also precede densification, but they are not needed if air classification or screening removes the metal.

The feedstock to a pellet mill is generally size reduced light fraction from an air classifier or size reduced screened oversize light fraction. The bulk density is 30 to 80 kg/m^3, and the moisture content must be in the range of 10 to 30 percent although 15 to 20 percent is generally desirable. The pellets essentially retain the chemical properties of the feedstock although the temperature at the wall of a die may be sufficient to effect localized pyrolysis. The size, shape, and bulk density of the material changes as has already been described. Also, some moisture is lost due to vaporization.

It should be noted that the models of the unit processes preceding densification have as an output a matrix of material characteristics.

The matrix information that is pertinent to densification is the maximum particle size. If the size is too large, the pellet mill cannot handle the material and therefore pellet formation cannot occur. The material composition of the output is similar to that of the input but is somewhat meaningless except as an input to heating value and ash calculations because the individual particles, being compacted into a pellet with many other particles, can no longer act as individual materials. Likewise, the size of the individual particles becomes meaningless upon densification, the size of the pellets being the parameter of primary importance.

The mass of solids entering and leaving the pellet mill is not affected by the operating variables. The mass of solids in the pellets is usually slightly less than the solids in the feedstock. Negligible amounts of dust are lost. The amount is determined by the pressure of extrusion, the moisture content of the feedstock, the cohesiveness of the feedstock, and possibly several other factors such as the shape of the die entrance, the temperature of the material during extrusion, and the rate and extent of moisture removal after extrusion. The density and bulk weight of the product, as well as the energy required to produce it, are dependent on the pressure of extrusion. The pressure depends upon the dimensions of the die and on certain properties of the feedstock, the most important of which is the moisture content.

A block diagram showing the mass balance variables discussed in the preceding paragraphs is shown in Figure 5.1.

Material is generally conveyed to a pellet mill via belt or pneumatic conveyors. It enters a feed hopper and is fed to the densification chamber via one or more screw conveyors. Since pellet mills are readily subject to overload by surges of feedstock, some manufacturers include automatic flow controls with the mill.

The equations used in the model to describe the mass flows and properties of the product are primarily empirical. The relevant physical law employed is conservation of mass.

The calculation of the pellet density is the most complex of the calculations in the mass balance model. The density depends upon the pressure required to extrude the pellet. The pressure depends on the diameter and length of the die, the degree of deformation of the

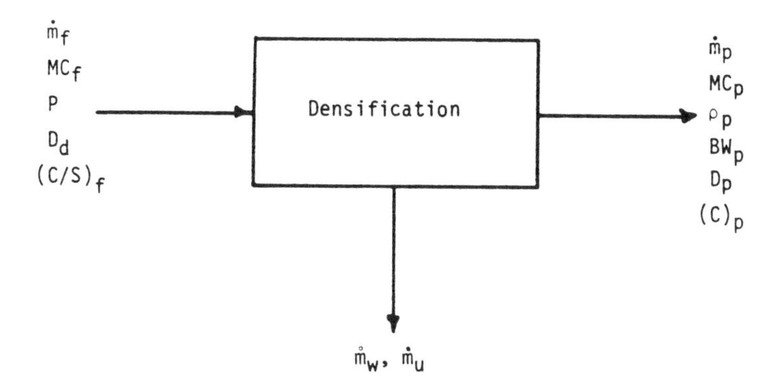

Nomenclature

BW = Bulk Density
D = Diameter
$\dot{m}$ = Mass Flowrate
MC = Moisture Content
P = Pressure
(C/S) = Composition-Size Matrix
(C) = Composition Matrix
ρ = Density

Subscripts

d = Die Opening
f = Feedstock
p = Pellet
u = Fines
w = Water

Figure 5.1. Densification Mass Balance Parameters

feedstock, friction, and other factors. These parameters are discussed fully in the section on the pellet mill energy model. Thus, pressure, which is an output of the energy model, is an input to the mass balance model.

Most of the data from which the mass balance model has been developed were the results of tests of several commercially available pellet mills. Pellet mills were used as the basis of the model because experience with them in MSW processing operations is much greater than that with briquetters or other machines used for producing d-RDF. Much of the mass balance model for pellet mills is also applicable to other types of densification equipment. However, the energy model that is presented later in this report is not applicable to machines other than pellet mills.

The equation for determining the density of pellets was derived from laboratory studies in which material was compressed under measured pressures in cylinders.

5.2 DEVELOPMENT OF MASS BALANCE MODEL

The inputs, outputs, and transfer functions for the mass balance model are summarized in Table 5.1. An explanation of each equation follows:

$$M_w = 0.01\ M_f \qquad (1)$$

The amount of evaporated water (M_w) depends on the moisture content of the feedstock, the temperature of the pellets upon exiting the mill, and the conditions under which the pellets cool. Pertinent conditions include the degree of ventilation, ambient humidity and temperature, and the degree of exposure of the hot pellets to the ambient air. No data have been found regarding the loss of water during densification. However, energy balance calculations indicate that under typical conditions the energy used to compress, deform, and force a pellet through a die is sufficient to vaporize about 0.02 g water/g extruded material. Since not all of the heat is used to vaporize water, 0.01 g vaporized water/g feedstock has been assumed to be a reasonable value for moisture loss.

$$M_p = 0.99\ M_f \qquad (2)$$

Table 5.1. Mass Balance Summary for Densification

Inputs	Equation	Outputs
1. Mass of feedstock (M_f)	$M_w = 0.01\ M_f$	Mass of evaporated water (M_w)
2. Mass of feedstock (M_f)	$M_p = 0.99\ M_f$	Mass of pellets (M_p)
3. Moisture content of feedstock (MC_f)	$MC_p = 1/0.99\ (MC_f - 0.01)$	Moisture content of pellets (MC_p)
4. Component-size matrix for feedstock $(x_{ij})_f$	$(x_i)_p = 0.99 \left(\sum_{i=1}^{5} x_{ij_f} \right)$	Component matrix for pellets $(x_i)_p$
5. Die hole smaller diameters (D_h)	$D_p = D_h$	Pellet diameter (D_p)
6. Extrusion pressure (P) [kPa] (from energy model)	$\rho = 3.56 + 1.33 \ln (P-103)$	Pellet density (ρ) [g/cm^3]
7. Pellet density (ρ) [g/cm^3]	$BW = 500\ \rho$	Bulk density of pellets (BW) [kg/m^3]

The mass of pellets is equal to the mass of feedstock minus the mass of evaporated water.

$$MC_p = 1/0.99\ (MC_f - 0.01) \tag{3}$$

This equation relates the moisture content of the pellets to the moisture content of the feedstock. It follows from the estimation that

$$M_w = 0.01\ M_f$$

$$(x_i)_p = 0.99\left(\sum_{j=1}^{5} x_{ij_f}\right) \tag{4}$$

This operation converts the component/size distribution of the feedstock to a component distribution of the pellets. The mass of the i'th component in the pellets is equal to 99 percent of the mass of the i'th component in the feedstock. The remaining one percent is vaporized water. It is assumed that the loss of vaporized water from each component is proportional to the mass of that component in the feedstock. No data was found in the study to confirm or refute this assumption.

$$D_p = D_h \tag{5}$$

The diameter of a pellet is essentially equal to the diameter of the hole in the die through which it is extruded. There often is a small amount of rebound or swelling after extrusion, but the amount is trivially small in most cases.

$$\rho = 3.56 + 1.33\ \ln\ (P-103) \tag{6}$$

where the units of the quantities are defined as: (1) ρ in g/cm^3; and (2) P in kPa_2.

This equation relates the density of a pellet to the pressure applied to form the pellet. The model for the pressure required to extrude a pellet is developed in the section on the energy model for densification. The relationship between pressure and density in the field of soil mechanics is known to be a logarithmic function. Ruf [1] in working with shredded refuse at pressures of up to 1400 kPa bars found the relationship to be of the form,

$$\rho = A + B\ \ln\ (P + C) \tag{7}$$

where:

P = pressure; and

A, B, and C are constants depending upon the properties of the material.

The reliability of the equation was checked using energy data obtained during a study conducted by the University of California (U.C.) [2] and by inferring, from basic fundamentals, that the marginal increase in specific energy is proportional to the corresponding marginal increase in pressure. The relation for pressure is discussed more fully in the section on the energy model.

The values of the constants A, B, and C derived by Ruf [1] do not correlate well with the U.C. data [2]. Reed, et al [3], however, in compressing sawdust to pressures of up to 70,000 kPa in laboratory test cylinders obtained data that fit the form of equation derived by Ruf when the constants A, B, and C are 3.56, 1.33, and 103 respectively when the dimensions of density are grams per cubic centimeter and the pressure is given in kilo pascals. Data from U.C. conform to Eq. 7 when the values of the constants are those cited by Reed. Since the equation predicts the results obtained from two independent sources for $0.9 < \rho < 1.3$ in which two different methods of compression are used, it is taken to be adequate for RDF densification.

$$BW = 500\ \rho \qquad (8)$$

Since the bulk density is, by definition, the average density of a group of particles times the fraction of a given volume that is actually occupied by particles, it follows that,

$$BW = (1\text{-}VR)\ \rho \qquad (9)$$

where:

VR = void ratio; and

BW and ρ are in consistent units.

Furthermore, particles of uniform size and shape and of a given arrangement (such as loose and randomly packed) do not undergo a change in bulk density if all of the particles undergo a uniform change in size. Therefore, if the amount of loose fines in a mixture of RDF pellets is assumed to be negligible, and the pellets are packed in a loose and random arrangement, any group of pellets should have a constant voids ratio and conform to the equation,

$$BW = A\rho \tag{10}$$

where A is a constant.

CRS has measured the bulk density of pellets to be 450 to 600 kg/m^3 [4,5]. Unfortunately, corresponding pellet densities were not reported. However, RDF pellets generally have densities in the range of 0.9 to 1.3 g/cm^3; and the ratio of the pellet densities in the cases where the bulk densities are 450 and 600 kg/m^3 is 1.33. It follows that the void ratio is 50 to 55 percent and that 'A' in the above equation has a value in the range of 450 to 500 when the units are kg/m^3 for bulk density and g/cm^3 for pellet density.

Wiles [6] reports data from the National Center for Resource Recovery (NCRR) that yields a value of A of 530 to 570. These higher values may have been due to the inclusion of a significant amount of fines in the samples. For the purposes of this model, the value of A is taken to be 500.

No test data are available for performing a comparative mass input/output analysis. According to the mass balance model, mass output is 99 percent of the input value, the loss being that of moisture vaporization as a consequence of the extrusion process.

It is convenient to discuss the pellet density predictive capability of the densification model in the energy model discussion inasmuch as pellet density, extrusion pressure, and energy requirements are intimately related. Consequently, an example is presented in the energy modeling section to illustrate the interrelationship among the variables and to compare predicted and actual values.

5.3 ENERGY REQUIREMENTS

5.3.1 Introduction

The specific energy consumption of densification is given by the relation,

$$E_s = \frac{\dot{W}}{\dot{m}} \tag{11}$$

where:

E_s = specific energy (kWh/Mg);

$\dot{W}$ = power consumption (kW); and

$\dot{m}$ = mass throughput rate (Mg/hr).

The mass throughput rate is selected by the plant designer.

Typically, a pellet mill rated at 4.5 Mg/hr has a main drive motor rated at 187 kW (250 hp). In practice, the mill may only achieve continuous throughput rates of about 3 to 4 Mg/hr. Also, during steady-state operation, the experience of CRS has been that in general the motor is substantially underutilized. Following is a method for determining the actual power that is derived from fundamental theory and from empirical data.

The power to the main drive motor of a pellet mill ($\dot{W}$) is subject to the following factors:

1. motor inefficiency ($\dot{W}_E$)
2. freewheeling power to turn the die of the pellet mill and the feeding mechanisms ($\dot{W}_{FW}$)
3. marginal power ($\dot{W}_M$) which is the power beyond the start-up power required for the pellet flow, and which comprises three factors:
   a) frictional power ($\dot{W}_F$) which is required to overcome the force of friction between a pellet and the die wall
   b) deformation power ($\dot{W}_D$) which is the power required to deform a compressed mass of a given cross-sectional area to a smaller cross-sectional area
   c) compression power ($\dot{W}_C$) which is the power required to compress a loosely packed mass such as fluff RDF into a densely packed mass
4. pellet start-up power ($\dot{W}_S$) which is the power required to initiate the flow of material through a die

5.3.2 Motor Inefficiency

The efficiency of an electric motor depends on the design of the motor and on the load. A fully loaded motor (i.e., one operating at its rated load) generally has an efficiency exceeding 90 percent, while an unloaded motor has an efficiency of zero percent. For the purposes of this model, 3.5 percent of the motor's rated power is assumed to be lost as heat due to the motor inefficiency. This value could correspond to a motor operating at about 30 percent of its rated load and an efficiency of 90 percent, or to a motor operating at 20 percent load and 85 percent efficiency. These values are in rough agreement with the characteristics of typical induction motors rated at 100 to 500 hp and with the typical operating loads of pellet mills tested by U.C. [2] and NCRR [7]. Thus,

$$\dot{W}_E = 0.035\ \dot{W}_R \tag{12}$$

where:

$\dot{W}_E$ = power loss due to motor inefficiency (kW); and
$\dot{W}_R$ = rated power of pellet mill main drive motor (kW).

5.3.3 Freewheeling Power

In the U.C. studies [2], 5 hp was required to turn the feed screws and the pellet mill die. This was about seven percent of the rated power of the drive motor (75 hp). In the model, the freewheeling power is taken to be,

$$\dot{W}_{FW} = 0.07\ \dot{W}_R \tag{13}$$

5.3.4 Pellet Start-Up Power

Data from both NCRR [7] and the U.C. [2] indicate that very small mass throughputs require a much higher marginal specific energy than do higher throughputs. This start-up phenomenon appears to be well pronounced at flows of up to 0.1 Mg/hr in the U.C. study.

The start-up phenomenon may be due to several factors, among which are the following:

1. Initiating the flow of a pellet requires that static friction be overcome. The frictional force resisting the flow of a pellet decreases once flow is initiated and kinetic friction replaces static friction.
2. When the layer of material being forced into a die becomes small in comparison to the diameter of the die into which it is flowing, the material begins to flow in a direction more nearly perpendicular to, rather than parallel to, the axis of the die opening. This change in the direction of the flow lines results in a dramatic increase in the pressure required to maintain flow.

Both of these phenomena were observed in bench-scale, single-die tests at the University of California.

From the NCRR [7] and U.C. [2] studies with commercial pellet mills, the following empirical equation for the start-up power has been derived:

$$\dot{W}_S = 0.004\ A \tag{14}$$

where:

$\dot{W}_S$ = start-up power (kW); and
A = area of inner die surface (cm^2).

5.3.5 Marginal Power

The marginal specific power (i.e., increase in power per additional unit of throughput) is fairly constant. The three components of marginal power (i.e., friction, deformation, and compression) are described individually below.

5.3.5.1 Friction

When the pressure is the same in all directions, the frictional force resisting the flow of a moving solid material through a cylindrical die opening is,

$$F_F = F_0 \exp\left(4\ \mu\ \frac{L}{D}\right) \tag{15}$$

where:

F_F = frictional force;
F_0 = a constant having units of force;
μ = coefficient of kinetic friction;
L = length of die; and
D = bore of the die openings.

The force terms (F_F and F_0) can be converted to pressure (P_F and P_0) by dividing by the cross-sectional area of the die openings. In the U.C. single-die study [2], the coefficient of kinetic friction was found to be 0.1 and P_0 was roughly 1000 kPa. Thus,

$$P_F = 1000 \exp\left(0.4 \frac{L}{D}\right) \tag{16}$$

where:

P_F is the pressure required to overcome kinetic friction and has units of kPa; and

L and D are in consistent units and are inputs to the model.

The frictional work is

$$W_F = \int P_F dV \tag{17}$$

where:

V = volume of extruded material.

Since Eq. 16 shows the pressure P_F to be constant for a given die the frictional work, power, and specific energy are,

$$W_F = P_F V \tag{18a}$$

$$\dot{W}_F = P_F Q \tag{18b}$$

where:

Q = volume flowrate.

$$E_F = \frac{P_F}{\rho} \tag{18c}$$

where:

ρ = density of the extruded pellets.

5.3.5.2 Deformation

The deformation pressure (P_D) is the pressure required to deform a cylinder of cross-sectional area 'A' to a cylinder of cross-sectional area 'a'. For a crystalline material in which plane sections remain plane during deformation, the theoretical deformation pressure is,

$$P_D = Y \ln \frac{A}{a} \tag{19}$$

where:

Y = yield stress.

Much of the simplicity of this equation (which is used primarily in the field of metallurgy) is lost when the material in question has a highly variable yield stress. The yield stress of a mass of RDF depends on the size and composition of the particles, the moisture content, the density of the mass, and the orientation of the forces acting on the mass. Nevertheless, Eq. 19 serves as a useful starting point in the prediction of the deformation pressure.

The U.C. single-die study [2] showed that, at a moisture content of 15 percent, the deformation pressure could be predicted with a reasonable degree of accuracy by Eq. 19 when Y equalled 24,000 kPa (3500 psi).

The U.C. pellet mill experiments and single-die experiments both showed that the effect of moisture content on pellet density was approximately

$$\frac{d\rho}{dMC} = -0.02 \tag{20}$$

where the units of ρ and MC are g/cm^3 and percent, respectively.

As was explained previously in the section on the mass balance for densification, the density of a pellet is related to the pressure required to form it by

$$\rho = 0.133 \ln (0.14P - 1500) \tag{21}$$

where ρ and P have units of g/cm^3 and kPa, respectively.

Thus,

$$\frac{d\rho}{dP} = \frac{0.133\ (0.14)}{0.14P - 1500} \tag{22}$$

Noting that the total pressure (P) required to extrude a pellet is equal to the sum of the pressure required to overcome friction (P_F) (which is assumed to be independent of the moisture content) and the pressure to deform the material (P_D), and combining Eqs. 21 and 22 yields,

$$\frac{dP_D}{dMC} = 0.143\ (11{,}280 - P_D - P_F) \tag{23}$$

Integration yields

$$-\ln(11{,}280 - P_D - P_F) = 0.143MC + C \tag{24}$$

where C is a constant of integration.

Evaluating C at a moisture content of 20 percent and a deformation pressure of 24,000 ln A/a yields

$$C = -\ln(11{,}280 - 24{,}000 \ln A/a - P_F) - 2.14 \qquad (25)$$

It follows that,

$$P_D = (24{,}000 \ln A/a) \exp [0.143(15\text{-}MC)] + (11{,}280 - P_F) \left\{1 - \exp [0.143(15\text{-}MC)]\right\} \qquad (26)$$

Eq. 26 indicates that the deformation pressure depends only on the area reduction ratio (A/a), the moisture content of the feedstock, and the frictional pressure. The frictional pressure affects the hardness or density of the pellets, thus affecting the yield stress. The area reduction ratio in a commercial pellet mill is the area of the inner surface of the die divided by the sum of the cross-sectional areas of all of the openings in the die at the outer surface of the die (i.e., where the pellets leave the die).

Intuitively, one would expect that the composition and the size distribution of the feedstock would affect the deformation pressure. However, there is no data that enables a quantification of the effect. Within the normally encountered variations in the characteristics of RDF feedstock, the model assumes there is no significant effect of the characteristics on the deformation pressure.

The deformation work, power, and specific energy are given by the following relations, respectively:

$$W_D = P_D V \qquad (27a)$$

$$\dot{W}_D = P_D Q \qquad (27b)$$

$$E_D = \frac{P_D}{\rho} \qquad (27c)$$

5.3.5.3 Total Pressure

The total pressure of extrusion is given by

$$P = P_D + P_F \qquad (28)$$

The pressure is an input to the mass balance model, and is used in determining the density of the product.

5.3.5.4 Compression

Before material can be deformed at the entrance to a die opening or pushed through the opening, it must be compressed from a loose fluff RDF state to a compacted state at pressure P (i.e., $P_D + P_F$). The work required for compression is

$$W_C = \int_{V_i}^{V_f} PdV \tag{29}$$

From Eq. 21, the pressure is related to density,

$$P = 7.1 \exp(7.52\rho) + 10{,}700 \tag{30}$$

This Eq. 30 is not easily integrated by analytic methods. Furthermore, it was developed for values of ρ of 0.9 g/cm^3 and greater.

Ruf [1] determined that for shredded MSW, the pressure and density were predicted by,

$$P = 7 \exp\left(\frac{\rho + 0.2}{0.15}\right) - 42 \tag{31}$$

This equation is valid up to a pressure of about 1300 kPa and a density of about 0.6 g/cm^3.

To determine the work and power of compression, one may integrate Eq. 29 using stepwise numerical methods and assume that Eq. 32 is valid up to a density of 0.7 while Eq. 31 is valid for densities of 0.8 and higher. The results of the integration are given in Table 5.2. The cumulative specific energy for densities in the range of 0.8 to 1.4 g/cm^3 is approximated by

$$E_C = 240\, e^{3.4\rho} \tag{32}$$

where:

E_C = specific energy of compression (kJ/Mg).

The power of compression is given by

$$\dot{W}_C = E_C \dot{m} \tag{33}$$

Table 5.2. Stepwise Integration of Equations 31 and 32

(1) Density (g/cm^3)	(2) Equation Used	(3) Pressure (kPa)	(4) Marginal Specific Energy (kWh/Mg)	(5) Cumulative Specific Energy kWh/Mg	(kJ/Mg)
0.1	18	10	0[a]	0	(0)
0.2	18	59	0.05	0.05	(180)
0.3	18	154	0.05	0.10	(360)
0.4	18	340	0.06	0.16	(576)
0.5	18	702	0.07	0.23	(828)
0.6	18	1,408	0.10	0.33	(1,188)
0.7	18	2,782	0.14	0.47	(1,692)
0.8	17	13,370	0.40	0.87	(3,132)
0.9	17	16,587	0.58	1.45	(5,220)
1.0	17	23,412	0.62	2.07	(7,452)
1.1	17	37,889	0.77	2.84	(10,224)
1.2	17	68,599	1.12	3.96	(14,256)
1.3	17	133,743	1.80	5.76	(20,736)
1.4	17	271,927	3.10	8.86	(31,896)

[a]The feedstock has a density of about 0.1 g/cm^3 so no work is required to attain this density.

5.3.6 Example

A summary of the methodology for determining the work for densification is presented in Table 5.3. Table 5.4 illustrates the use of the energy model. The specific energy predicted by the model is about 10 to 20 percent less than was obtained by the University of California in measurements conducted on a 75 hp pellet mill. The underestimation of the specific energy is probably appropriate inasmuch as the larger mills (250 hp) typically used in commercial operations can be expected to operate more efficiently than the 75 hp mill used to develop the energy model. Furthermore, the error decreases as the throughput increases.

5.4 REFERENCES

1. Ruf, J., Particle Size Specturm and Compressibility of Raw and Shredded Municipal Solid Waste, Ph.D. thesis, University of Florida, 1974.

2. Trezek, G.J., G.M. Savage, and D.B. Jones, Fundamental Considerations for Preparing Densified Refuse Derived Fuel, U.S. EPA, Grant No. R-805414-010, 1981.

3. Reed, T.B., G.J. Trezek, and L.F. Diaz, "Biomass Densification Energy Requirements," ACS Symp. Series, No. 130, Thermal Conversion of Solid Wastes and Biomass, Edited by Jones and Radding, 1980.

4. Tuck, J.K, and G.M. Savage, Densified Refuse-Derived Fuel Characteristics, Test Methods, and Specifications for Medium Capacity Boiler Facilities, Report to the Department of the Navy, Civil Engineering Laboratory, Port Hueneme, California, Sept. 1981.

5. Cal Recovery Systems, Inc., Bulk Density Measurements of Selected Fractions of Processed MSW, Report to the American Society for Testing and Materials, March 1981.

6. Wiles, C.C., "The Production and Use of a Densified Refuse Derived Fuel," Fifth Annual Research Symp., Land Disposal and Resource Recovery, U.S. EPA, Orlando, Florida, March 1979.

7. National Center for Resource Recovery, "Summary of Project and Findings - Preparation of dRDF on a Pilot Scale," June 1977.

Table 5.3. Densification Energy Model Summary

Input to Model	Eqn. No.	Equation	Output
L,D	5	$P_F = 1000\exp[0.4\,(L/D)]$	P_F [kPa]
A,a,MC	13	$P_D = (24{,}000 \ln A/a)\exp[0.143(15-MC)] + (11{,}280 - P_F)[1-\exp[0.143(15-MC)]$	P_D [kPa]
	15	$P = P_D + P_F$	P [kPa]
	10	$\rho = 0.133 \ln[0.14P - 1500]$	ρ [g/cm^3]
	7	$E_F = P_F/\rho$	E_F [kJ/Mg]
	14	$E_D = P_D/\rho$	E_D [kJ/Mg]
	32	$E_C = 240\,e^{3.4\rho}$	E_C [kJ/Mg]
	--	$E_M = (E_F + E_D + E_C)/3600$	E_M [kWh/Mg]
	4	$\dot{W}_S = 0.004\,A$	$\dot{W}_S$ [kW]
$\dot{W}_R$	3	$\dot{W}_{FW} = 0.07\,\dot{W}_R$	$\dot{W}_{FW}$ [kW]
	2	$\dot{W}_E = 0.035\,\dot{W}_R$	$\dot{W}_E$ [kW]
$\dot{m}$	--	$\dot{W} = \dot{W}_E + \dot{W}_{FW} + \dot{W}_S + E_M\dot{m}$	$\dot{W}$ [kW]
	--	$E_O = \dot{W}/\dot{m}$	E [kWh/Mg]

Inputs

- L = Die opening length [cm]
- D = Die opening diameter [cm]
- A = Die internal surface area [cm^2]
- a = Total die opening cross sectional area [cm^2]
- MC = Feedstock moisture content [%]
- $\dot{W}_R$ = Drive motor rated power [kW]
- $\dot{m}$ = Mass throughput [Mg/hr]

Outputs

- P_F = Pressure to counteract static friction [kPa]
- P_D = Deformation pressure [kPa]
- P = Total extrusion pressure [kPa]
- ρ = Pellet density [g/cm^3]
- E_F = Frictional specific energy [kJ/Mg]
- E_D = Deformation specific energy [kJ/Mg]
- E_C = Compression specific energy [kJ/Mg]
- E_M = Marginal specific energy [kWh/Mg]
- $\dot{W}_S$ = Start-up power [kW]
- $\dot{W}_{FW}$ = Freewheeling power [kW]
- $\dot{W}_E$ = Motor power loss [kW]
- $\dot{W}$ = Total densification power [kW]
- E_O = Total densification specific energy [kWh/Mg]

Table 5.4. Densification Energy Model Example

Inputs

Drive motor rated power = 75 hp = 56 kW

Die opening diameter = 1.0 inch = 2.54 cm

Die opening length = 5 inches = 12.7 cm

Inner die surface area = 314 in^2 = 2026 cm^2

Total die opening area = 86.4 in^2 = 557 cm^2

Feedstock moisture content = 15 percent

$\dot{m}$ ranges from 0.1 Mg/hr to 0.7 Mg/hr

Feedstock: Screened light fraction

Outputs

P_F = 7,390 kPa (1070 psi)

P_D = 30,970 kPa (4490 psi)

P = 38,360 kPa (5560 psi)

ρ = 1.10 g/cm^3

E_F = 6,720 kJ/Mg; E_D = 28,150 kJ/Mg: E_C = 10,220 kJ/Mg

E_M = 45,090 kJ/Mg (12.5 kWh/Mg)

$\dot{W}_S$ = 8.1 kW; $\dot{W}_{FW}$ = 3.9 kW; W_E = 2.0 kW

$\dot{W} = 14.0 + 12.5\,\dot{m}$

$E = 12.5 + 14.0/\dot{m}$

Throughput (Mg/hr)	0.1	0.2	0.3	0.4	0.5	0.6	0.7
Specific Energy (Model)	152.5	82.5	59.2	47.5	40.5	35.8	32.5
Specific Energy (from U.C. study)	190.0	92.0	71.0	58.0	49.0	42.0	36.6

6. Conveying Model

6.1 BACKGROUND

The principal means of transporting material among unit operations in resource recovery facilities is the use of mechanical conveyors. The types of mechanical conveyors used are primarily belt and apron conveyors. The type of conveyor used depends upon the material and upon the unit operation served by the conveyors. Raw MSW is typically transported by an apron conveyor, while rubber belted conveyors generally are used for size reduced material.

The key operating parameters in the design of conveyors and in the development of the conveyor model include mass flowrate, material composition, and its particle size distribution.

Shown in Figure 6.1 is a block diagram of a conveyor with the relevant inputs and outputs. The subscript 'x' refers to individual components (i.e., paper, glass, etc.) while the subscripts i, o, and s denote input, process output, and spillage, respectively.

The conveyor model was developed to a large extent from information and test data presented by Khan, *et al* [1]. Khan and his co-workers studied the spillage of several processed waste fractions from horizontal and inclined belt conveyors.

6.2. DEVELOPMENT OF MASS BALANCE MODEL

The conveyor is modeled as a split stream process with the input material being divided into two streams, spillage and process output. Inputs, transfer functions, and outputs are summarized in Table 6.1. The inputs include total mass flowrate, $\dot{m}_i$, component mass fraction, mf_x, component size distribution, $\underline{X}_x$, component mass flowrate, $\underline{\dot{m}}_{x,i}$, spillage factor, $\underline{\sigma}_x$, and conveyor length, L. Component mass flowrates are expressed as matrices with respect to size distribution and are broken down by component. The spillage factor is also size and material specific. It is expressed as a fraction of component mass input per unit length of conveyor. Therefore, knowing the input mass flowrate, conveyor length,

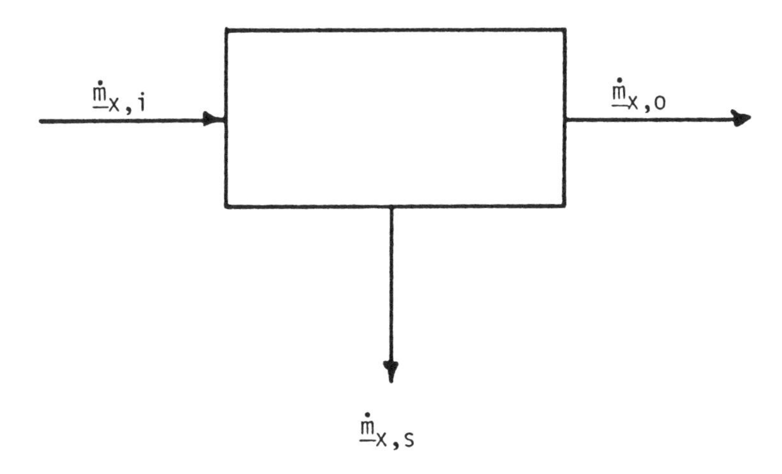

Figure 6.1. Conveyor Flow Diagram

Table 6.1 Conveyor Transfer Relations

Inputs	Transfer Equation	Output
a) $\dot{m}_i$; mf_x; $\underline{X}_{\underline{x},i}$	$\underline{\dot{m}}_{\underline{x},i} = mf_x \times \dot{m}_i \times \underline{X}_{\underline{x},i}$	$\underline{\dot{m}}_{\underline{x},i}$
b) $\underline{\dot{m}}_{\underline{x},i}$; L; $\underline{\sigma}_{\underline{x}}$	$\underline{\dot{m}}_{\underline{x},s} = \underline{\sigma}_{\underline{x}}\ L\ \underline{\dot{m}}_{\underline{x},i}$	$\underline{\dot{m}}_{\underline{x},s}$
c) $\dot{m}_{x,i}$; L; $\underline{\sigma}_{\underline{x}}$	$\underline{\dot{m}}_{\underline{x},o} = (1-\underline{\sigma}_{\underline{x}})\ L\ \underline{\dot{m}}_{\underline{x},i}$	$\underline{\dot{m}}_{\underline{x},o}$

$\underline{\dot{m}}_{\underline{x},i}$ = Input mass flowrate of component x by size class conveyor.

L = Conveyor length.

$\underline{\sigma}_{\underline{x}}$ = Spillage by size class of component x per unit mass input flow per unit length of conveyor.

$\underline{\dot{m}}_{\underline{x},s}$ = Mass spillage rate of component x.

$\underline{\dot{m}}_{\underline{x},o}$ = Mass flowrate of component x by size class output from the conveyor.

$\dot{m}_i$ = Total input mass flowrate.

mf_x = Mass fraction of component x.

$\underline{X}_{\underline{x},i}$ = Size distribution.

and spillage factor, the spillage rate and process output flowrate can be determined. The following example demonstrates how the mass balance model predicts the spillage rate of a given component.

6.2.1 Example

The following example illustrates the use of the conveyor mass balance model. First, the inputs to the model are defined:

1. Total mass input flowrate:
   $\dot{m}_i$ = 20 Mg/hr
2. Component (x) = glass (gl)
3. $mf_{gl,i} = 0.1$
4. j = number of size classes
   = 5
5. Weight Percent / Size (mm)

$$\underline{X}_{gl,i} = \begin{bmatrix} 0.30 \\ 0.27 \\ 0.23 \\ 0.06 \\ 0.14 \end{bmatrix} \quad \begin{matrix} +9.53 \\ -9.53 \; +6.35 \\ -6.35 \; +2.82 \\ -2.82 \; +1.42 \\ -1.42 \end{matrix}$$

6. $\underline{\sigma}_{x,j} = [0.4 \quad 0.4 \quad 0.4 \quad 0.4 \quad 0.4] \; \frac{(kg)}{(Mg \cdot m)}$

Size class (mm)	+9.53	-9.53 +6.35	-6.35 +2.82	-2.82 +1.42	-1.42

7. L = 10 m

The outputs are computed as follows:

1. Input mass flowrate of glass ($\underline{\dot{m}}_{gl,i}$)

$$\underline{\dot{m}}_{gl,i} = mf_{gl,i} \times \dot{m}_i \times \underline{X}_{gl,i}$$

$$= (0.1)\,(20) \times \begin{bmatrix} 0.30 \\ 0.27 \\ 0.23 \\ 0.06 \\ 0.14 \end{bmatrix} \text{ Mg/hr}$$

$$= \begin{bmatrix} 0.60 \\ 0.54 \\ 0.46 \\ 0.12 \\ 0.28 \end{bmatrix} \text{ Mg/hr}$$

2. Cumulative mass flowrate of glass in the input stream ($\dot{m}_{gl,i}$)

$$\dot{m}_{gl,i} = \sum_{j=1}^{n} (\dot{m}_{gl,i})_j$$

$$= 0.60 + 0.54 + 0.46 + 0.12 + 0.28 \text{ Mg/hr}$$

$$= 2.0 \text{ Mg/hr}$$

3. Glass spillage rate ($\underline{\dot{m}}_{x,s}$)

$$\underline{\dot{m}}_{x,s} = L \times \underline{\sigma}_{gl} \times \underline{\dot{m}}_{gl,i}$$

$$= (10) \times [0.4 \quad 0.4 \quad 0.4 \quad 0.4 \quad 0.4] \times \begin{bmatrix} 0.60 \\ 0.54 \\ 0.46 \\ 0.12 \\ 0.28 \end{bmatrix} \text{ kg/hr}$$

$$= \begin{bmatrix} 2.40 \\ 2.16 \\ 1.84 \\ 0.48 \\ 1.12 \end{bmatrix} \text{ kg/hr}$$

4. Cumulative glass spillage rate

$$\dot{m}_{gl,s} = \sum_{j=1}^{5} (\dot{m}_{gl,s})_j$$

$$= 2.40 + 2.16 + 1.84 + 0.48 + 1.12 \text{ kg/hr}$$

$$= 8 \text{ kg/hr}$$

5. Size distribution of spilled glass ($\underline{X}_{gl,s}$)

$$\underline{X}_{gl,s} = \frac{\underline{\dot{m}}_{gl,s}}{\dot{m}_{gl,s}}$$

$$= \begin{bmatrix} 2.40 \\ 2.16 \\ 1.84 \\ 0.48 \\ 1.12 \end{bmatrix} \text{ kg/hr}$$

---

8 kg/hr

$$= \begin{bmatrix} 0.30 \\ 0.27 \\ 0.23 \\ 0.06 \\ 0.14 \end{bmatrix}$$

6.3 ENERGY REQUIREMENTS

The power required to operate a conveyor is the sum of the power required to move an empty belt, to convey material horizontally, and to lift material. All three components of the power can be derived from basic physical laws for both apron conveyors and belt conveyors. However, a knowledge of the value of the coefficient of friction between sliding parts is required to determine the first two components, and the value may vary considerably for various conveyor widths and lengths. Furthermore, stretching of belts and other factors complicate the calculations. Empirical equations have been developed for the power requirements of belt conveyors. Theoretical equations, with assumed values of apron weights and coefficients of friction, are used here for apron conveyors. Both methods are based on models given by Henderson and Perry [2] and by Marks [3].

6.3.1 Belt Conveyors

The three components of the power to drive a belt conveyor are,

$$P_1 = 1.5\ v\ (A + BL) \tag{1}$$

$$P_2 = 0.0082\ \dot{m}\ (0.48 + 0.01\ L) \tag{2}$$

$$P_3 = 0.0027\ H\ \dot{m} \tag{3}$$

where:

P_1 = power to drive empty belt (kW);
P_2 = power to convey material on level (kW);
P_3 = power to lift material (kW);
v = belt speed (m/s);
A,B = coefficients based on belt width;
$\dot{m}$ = mass flowrate (Mg/hr);
L = belt length (m); and
H = height of lift (m).

The mass flowrate, belt length, and height of lift are all inputs to the conveyor model, the mass flowrate being derived from the mass balance of the equipment feeding the belt and the last two variables being chosen by the designer.

For purposes of this model, a typical belt speed of 0.8 m/s can be used. The coefficient 'A' can be approximated by,

$$A = 0.60\ W \tag{4}$$

where:

W = belt width (m).

This approximation is derived from tabulated values given in Ref. 2. The coefficient 'B' can likewise be approximated as,

$$B = 0.013\ W \tag{5}$$

The belt width is

$$W = 0.278\ \dot{m}/(\rho v d) \tag{6}$$

where:

W = belt width (m);
$\dot{m}$ = mass flowrate (Mg/hr);
ρ = material density (kg/m^3);
v = belt speed (m/s); and
d = average depth of material over the width of the belt (m).

The density is an input to the conveyor model, and the belt speed is assumed to be 0.8 m/s. The average depth can be taken to be 0.05 W. This value is a compromise between acceptable maximum depths given in

Ref. 2 (about 0.09 W) and depths derived from data provided by one manufacturer (about 0.02 W). It follows that

$$W = 2.6\ (\dot{m}/\rho)^{1/2} \tag{7}$$

The coefficients 'A' and 'B' can then be given as

$$A = 1.6\ (\dot{m}/\rho)^{1/2} \tag{8}$$

$$B = 0.034\ (\dot{m}/\rho)^{1/2} \tag{9}$$

Substituting these values into Eq. 1, letting the belt velocity be 0.8 m/s, and assuming a motor efficiency of 90 percent yields

$$P_{bc} = 2.1\ \left(\frac{\dot{m}}{\rho}\right)^{1/2} (1 + 0.021\ L) + 0.0044\ \dot{m}\ (1 + 0.021\ L + 0.69\ H) \tag{10}$$

where:

P_{bc} = power requirement of a belt conveyor (kW).

6.3.2 Apron Conveyors

The power required to move an apron conveyor is the product of frictional force resisting the motion and the velocity of the conveyor. The addition of material to the conveyor increases the power requirement by increasing the frictional force which is directly proportional to the weight of material on the conveyor. The power required to raise the material to any desired elevation is the same as for belt conveyors. Marks [3] gives the total force required for conveying as

$$F_{ac} = L\ (F + F_1)\ (\mu \cos\theta + \sin\theta) + FL\ (\mu \cos\theta - \sin\theta) \tag{11}$$

where:

F_{ac} = force to move the apron conveyor;
L = conveyor length;
F = weight of the conveyor per unit length;
F_1 = weight of the material on the conveyor per unit length;
μ = coefficient of friction between the conveyor and the channel in which it slides; and
θ = angle of inclination of the conveyor.

The coefficient of friction of rollers that are often used is about 0.1. The length is specified by the designer. The weight of the conveyor per unit of length is

$$F = 500\ W \tag{12}$$

where:

F = weight per length (N/m); and
W = conveyor width (m).

Eq. 12 is based on the weight of steel aprons supplied by one manufacturer. The width of the apron is specified by the designer. The weight of material on the belt per unit of length is equal to

$$F_1 = 0.28\ \frac{\dot{m}g}{v} \tag{13}$$

where:

F_1 = weight of material per length (N/m);
v = conveyor velocity; and
g = acceleration of gravity (9.8 m/s^2).

The velocity is selected by the designer. The angle of inclination is given by

$$\theta = \sin^{-1}\ (\frac{H}{L}) \tag{14}$$

The height of lift (H) is selected by the designer. Eq. 11 can be rewritten as

$$F_{ac} = (L^2 - H^2)^{1/2}\ (100\ W + 0.27\ \dot{m}/v) + 2.7\ H\ \dot{m}/v \tag{15}$$

The power, when the motor efficiency is 90 percent is

$$P = (L^2 - H^2)^{1/2}\ (0.11\ W\ V + 0.0003\ \dot{m}) + 0.003\ H\ \dot{m} \tag{16}$$

where:

P = power (kW);
L = conveyor length (m);

$\dot{m}$ = mass flowrate (Mg/hr);

H = lift (m); and

v = conveyor velocity (m/s).

6.4 REFERENCES

1. Khan, Z., M.L. Renard, and J. Campbell, Considerations in Selecting Conveyors for Solid Waste Applications, EPA-60/2-82-082, MERL, Cincinnati, Ohio, Sept. 1982.

2. Henderson, S.M. and R.L. Perry, Agricultural Process Engineering, Avi Publishing Co., Inc., Westport, Connecticut, 1976.

3. Marks, L.S., Standard Handbook for Mechanical Engineers, McGraw-Hill, Inc., New York, 1978.

Part III

Task 4 Report: Economic Models

Preface

The Task 4 Report presents the economic models for the unit operations described in the Task 3 Report, namely:

- Size Reduction
- Air Classification
- Trommel Screening
- Ferrous Separation
- Densification
- Conveying

The economic models are broken down in terms of capital, operating, and maintenance cost elements. In the report are presented the development of the cost models, background information, and governing relations.

1. Introduction

An economic model for the unit operations in a resource recovery system is presented herein. The report is organized according to cost component (i.e., capital, operation, maintenance, residue disposal, and revenue from the sale of materials) with the unit operations (size reduction, air classification, screening, ferrous material separation, densification, and conveying) discussed as sub-categories under each cost component. The reason for this method of organization is that the portion of each cost component that is attributable to each unit operation often is not determinable when the unit operation is considered in isolation from the rest of the system.

The capital and maintenance cost equations primarily are based on price estimates and quotes from equipment suppliers, from published data on existing resource recovery operations, and from information obtained during interviews with managers of existing operations. In a few cases, estimates of the labor required to maintain equipment are based on CRS's experience. A wage rate of $30/hr is assumed in these instances. Maintenance costs include both labor and supplies.

The base cost of the equipment (i.e., uninstalled) is adjusted by the amounts indicated in Table 1.1 to arrive at an installed cost.

The operating labor requirements were estimated on the basis of CRS's in-plant field test experience and engineering judgment. The energy requirements are an input to the economic model. They are determined from the mass and energy requirements reported in the Task 3 Report.

The revenue from the sale of recovered materials and the cost of disposing residue are derived from the mass of recovered material and of residue, both of which are calculated in the mass balance model and are, thus, inputs to the economic model.

Several design and site-specific cost inputs must be supplied by the user of this model. They are noted in the text as the need for them arises.

Table 1.1. Adjustments to Base Costs of Equipment

Cost Item	Equipment	Adjustment[a] (%)
Engineering services	All	10
Contractor overhead and profit	All	10
Instrumentation, controls, and other accessory equipment	All	20
Sales tax	All	5
Transportation	All	5
Foundations	Hammermill shredder	12
	Air classifier	5
	Trommel screen	5
	Ferrous magnet	15
	Pellet mill	5
	Conveyors	2
Installation	Hammermill shredder	9
	Air classifier	22
	Trommel screen	22
	Ferrous magnet	22
	Pellet mill	9
	Conveyors	30

[a] Adjustments are given as percentages of the uninstalled equipment cost.

The accuracy of the model is estimated to be as follows:

1. Capital costs - ±20 percent
2. Operating costs
    a. Labor - ±30 percent
    b. Energy - +30 percent, -10 percent
3. Maintenance costs - ±30 percent
4. Disposal costs - accuracy is limited by the accuracy of the mass balance model
5. Revenue - accuracy is limited by the accuracy of the mass balance model

The requirement for operating labor depends on the design and degree of automation of the system. The associated error is estimated to be within the range of ±30 percent. The maintenance cost data provided by plant managers and by equipment suppliers varied widely. Consequently, the maintenance costs estimated by the model are considered accurate within the range of ±30 percent.

An important limitation on the accuracy of the model derives from the assumption that equipment is operated at its rated capacity. The underutilization of equipment can result in a significantly higher unit cost of capital than what the model predicts. Other costs may also be affected to a smaller degree.

The model is valid for the following mass throughput rates:

1. Hammermill shredder - >20 Mg/hr
2. Air classifier - 20-80 Mg/hr
3. Trommel screen - 20-80 Mg/hr
4. Ferrous magnet
    a. Shredded MSW feedstock - 15-70 Mg/hr
    b. Air classified feedstock - 37-130 Mg/hr
5. Pellet mill - >4 Mg/hr
6. Conveyors
    a. Belt conveyors - >3 Mg/hr for material of 50 kg/m^3 density (a correspondingly higher limiting throughput exists for denser material)
    b. Apron conveyor - Belt width of 1.2-2.4 m

In cases where an upper limit to the throughput rate is given, the costs for higher rates may be estimated by assuming that the unit cost for the higher rate is equal to the unit cost for the maximum rate specified above. The limits on the throughput rates mainly arise from the trends evident in the capital cost for various sizes of equipment. Upper limits on the throughput rates occur for unit operations for which there are significant economies of scale within the given range of throughputs.

Costs that are not included in the model include:

1. Building and site improvement
2. Utility connections to the plant site
3. Supervisory, clerical, janitorial, and other indirect labor
4. Transportation of salable materials and residue except to the extent that the user of the model includes these costs in the unit price of recovered material and the unit cost of residue disposal
5. Loading raw MSW into the initial unit operation
6. Storage of MSW, residue, and recovered materials

2. Capital Costs

The models described herein represent the capital costs (in 1984 dollars) of refuse processing unit operations. The models for each unit operation (i.e., size reduction, etc.) are developed and reported in separate subsections. A summary of the relations for the capital costs of the various unit operations is given in Table 2.1.

2.1 SIZE REDUCTION

CRS [1] found the average price of several hammermills for primary and secondary size reduction to be $10,600/Mg/hr and $11,600/Mg/hr (1984 dollars) respectively. The mills considered were designed to shred 30 to 60 Mg/hr in the case of primary size reduction and 30 to 50 Mg/hr in secondary size reduction applications. Since there was substantially no economy of scale over those ranges, the prices can be considered valid at throughputs of 20 Mg/hr or more.

Including the adjustments given in Table 1.1, the installed capital costs become:

$$C_{ph} = 18{,}100\ \dot{m}_{ph} \tag{1}$$

$$C_{sh} = 19{,}800\ \dot{m}_{sh} \tag{2}$$

where:

C_{ph} = installed cost of a primary hammermill shredder ($);
C_{sh} = installed cost of a secondary hammermill shredder ($);
$\dot{m}_{ph}$ = rated capacity of the primary shredder (Mg/hr); and
$\dot{m}_{sh}$ = rated capacity of the secondary shredder (Mg/hr).

2.2 AIR CLASSIFICATION

The cost of air classification is more variable than the cost of most other refuse processing equipment. There are some basic differences in the design of air classifiers used in resource recovery operations

Table 2.1. Capital Costs of Unit Operations

Unit Operation	Type of Equipment	Cost ($)	Range of Throughputs (Mg/hr)
Primary Size Reduction	Hammermill Shredder Shear Shredder	$18{,}100\ \dot{m}$	> 20
Secondary Size Reduction	Hammermill Shredder	$19{,}800\ \dot{m}$	> 20
Air Classification	Vertical Air Classifier with Cyclone and Baghouse	$16{,}000\ \dot{m} - 100\ \dot{m}^2$	20 to 80
Screening	Trommel Screen Disc Screen	$8600\ \dot{m} - 13\ \dot{m}^2$	20 to 80
Ferrous Material Recovery	Multiple Cross-Belt Magnet	$71{,}900 + 540\ \dot{m}$ [a] $71{,}900 + 220\ \dot{m}$ [b]	15 to 70 37 to 175
Densification		$41{,}000\ \dot{m}$	> 4
Conveying	Belt Conveyor	$370[2.6(\frac{\dot{m}}{\rho})^{0.5}(100+1.1L)+L][1+0.2(\sin^{-1}\frac{H}{L})]$	< 3[c]
	Apron Conveyor	$1020[6.9+L][7+W][1+0.2(\sin^{-1}\frac{H}{L})]$	(d)

[a] Equation valid for shredded MSW.
[b] Equation valid for air classified heavy fraction.
[c] Lower limit of throughput rate is given for material of 50 kg/m^3 density. The lower limit would be higher for denser material.
[d] Valid for belt widths of 1.2-2.4 m and inclination angles of 0-0.8 rad.

(e.g., horizontal, vertical, vibrating systems, etc.) that give rise to the price variations. Furthermore, most air classifiers include extensive duct work for conveying both solids and air. The configuration of the duct work is generally site-specific.

Cost data collected by Midwest Research Institute (MRI) [2] indicate that the costs of two air classifiers designed to process similar amounts of refuse may vary by a factor of two or more. However, if the small experimental air classifiers that were studied are not considered in the development of a capital cost model, the cost data collected by MRI supports the following capital cost equation:

$$C_{ac} = 16{,}000\ \dot{m}_{ac} - 100\ \dot{m}_{ac}^2 \qquad (3)$$

where:

C_{ac} = installed cost of an air classifier with a cyclone and baghouse ($); and

$\dot{m}_{ac}$ = rated mass throughput rate of the air classifier (Mg/hr).

Eq. 3 is valid in the range of 20 to 80 Mg/hr.

2.3 TROMMEL SCREENING

Adjusting the average of costs obtained from two manufacturers of trommel screens by the factors given in Table 1.1 yields the following cost equation for trommel screens:

$$C_{ts} = 8600\ \dot{m}_{ts} - 13\ \dot{m}_{ts}^2 \qquad (4)$$

where:

C_{ts} = installed cost of trommel screen ($); and

$\dot{m}_{ts}$ = rated mass throughput rate of the trommel screen (Mg/hr).

Eq. 4 is valid in the range, $20 < \dot{m}_{ts} < 80$ (Mg/hr).

2.4 FERROUS MATERIAL RECOVERY

Ferrous material may be recovered by a single cross-belt magnet, a multiple cross-belt magnet, a rotating drum magnet, or other devices. A multiple cross-belt magnet is used as a basis for the model.

Cost information was obtained from a manufacturer of a multiple cross-belt (MCB) magnetic separation system specifically designed for MSW. Such equipment is used in several full-scale resource recovery operations. Adjusting the prices quoted by the manufacturer for MCB magnetic separation systems suitable for throughputs of 15 to 70 Mg of shredded MSW per hour yields,

$$C_{fs} = 73{,}100 + 550\ \dot{m}_{fe} \qquad (5)$$

where:

C_{fs} = installed cost of ferrous material recovery system for shredded MSW feedstock ($); and

$\dot{m}_{fe}$ = rated mass throughput rate of the magnet for shredded MSW feedstock (Mg/hr).

Other firms quoted prices for single cross-belt magnets that are about 30 to 50 percent of the price given above. This is reasonable inasmuch as the modeled system includes three magnets.

Eq. 5 is valid for shredded MSW. The capacity of a ferrous separation system is actually a function of volume rather than mass. Thus, rated mass throughput rate should vary linearly with the density of the feedstock. Since the heavy fraction from an air classifier is typically 2.5 times as dense as shredded MSW, the cost function for a ferrous material removal system when the heavier material is the feedstock is,

$$C_{fh} = 71{,}900 + 220\ \dot{m}_{fh} \qquad (6)$$

where:

C_{fh} = installed cost of ferrous material recovery system for air classified heavy fraction ($); and

$\dot{m}_{fh}$ = rated mass throughput rate of the magnet for air classified heavy fraction (Mg/hr).

Eq. 6 is valid in the range of 27 to 130 Mg/hr.

2.5 DENSIFICATION

Two manufacturers quoted prices in the range of $75,000 and $100,000 for pellet mills rated at 4.5 Mg/hr. Both machines, according to the manufacturers, can handle more than 6 Mg/hr of feedstock under conditions of uniform flow and ideal moisture content (15 to 20 percent). However, the costs must be interpreted in light of the fact that the pellet mills handling RDF never achieve the quoted values of capacity. For example, the pellet mills at the Baltimore County resource recovery plant have a throughput capacity of about 2.7 Mg/hr, although they are rated at 4.5 to 6.4 Mg/hr.

In the model, the throughput rate is taken to be 4 Mg/hr, the higher cost figure (i.e., $100,000) is used, and the total adjustment factor is 64 percent. Thus, for throughput rates of greater than 4 Mg/hr,

$$C_D = 41{,}000\ \dot{m}_D \tag{7}$$

where:

C_C = installed cost of pellet mill ($); and

$\dot{m}_D$ = mass throughput rate of pellet mill (Mg/hr).

2.6 CONVEYORS

Information from three manufacturers indicates that the cost of belt conveyors can be approximated by

$$C = [20{,}000\ W + 200\ (1 + 1.1\ W)\ L]\ [1 + 0.2\ (\sin^{-1} \frac{H}{L})] \tag{8}$$

where:

C = cost of belt conveyor ($);
W = belt width (m);
L = length of conveyor (one half of the belt length) (m);
H = vertical lift (m); and
$\sin^{-1}$ = angle (rad).

It was shown in the section on conveyor power that the appropriate belt width can be calculated as

$$W = 2.6\ (\dot{m}/\rho)^{1/2} \tag{9}$$

where:

$\dot{m}$ = mass flowrate (Mg/hr); and
ρ = density (kg/m^3).

Substituting for W in Eq. 8 and multiplying the base cost by 1.85 to account for installation yields the following cost equation for an installed belt conveyor,

$$C_{bc} = 370\ [2.6\ (\dot{m}/\rho)^{1/2}\ (100 + 1.1\ L) + L]\ [1 + 0.2\ (\sin^{-1} \frac{H}{L})] \quad (10)$$

Eq. 8 was developed from cost data obtained for belts of 0.6-1.2 m in width. It follows from Eq. 9 that the minimum throughput rate for which the model is valid depends on the density of the material. For a material with a density of 50 kg/m^3, the minimum throughput is about 3 Mg/hr.

The cost of apron conveyors supplied by one manufacturer is approximated by

$$C = [3800 + 550L]\ [7 + W]\ [1 + 0.2\ (\sin^{-1} \frac{H}{L})] \quad (11)$$

where:

C = cost of apron conveyor ($);
W = conveyor width (m);
L = conveyor length (m); and
H = lift (m).

Adjusting the base cost by 85 percent yields the installed cost,

$$C_{ap} = 1020\ [6.9 + L]\ [7 + W]\ [1 + 0.2\ (\sin^{-1} \frac{H}{L})] \quad (12)$$

Eq. 12 is valid for belt widths of 1.2 to 2.4 m and inclination angles of up to 0.8 rad.

2.7 REFERENCES

1. Savage, G.M., D.J. Lafrenz, D.B. Jones, and J.C. Glaub, Engineering Design Manual for Solid Waste Reduction Equipment, prepared for U.S. EPA under Contract No. 68-03-2972, 1982.

2. Hopkins, V., B.W. Simister, and G.M. Savage, Comparative Study of Air Classifiers, prepared for U.S. EPA under Contract No. 68-03-2730, 1981.

3. Operating Costs

The operating costs are composed of labor costs and energy costs. The unit processes discussed in this report use electrical energy only. Usually, no operating supplies are required. Materials and supplies required for maintenance are included in the maintenance cost model.

In the cost model the labor requirements are determined for the entire processing system rather than for each individual unit operation, except in the case of densification. The reason is that the labor required for a unit operation depends to a large extent on the configuration of the entire system. Furthermore, the allocation of labor among pieces of equipment is difficult to model. The allocation of labor for conveyors is one example.

The model for labor is valid for systems with throughput rates exceeding 10 Mg/hr and is designed to account for multiple processing lines. The model for energy costs requires inputs from the models for the energy requirements of the various unit operations.

3.1 LABOR

The unit cost of labor is given by the equation,

$$UCL = \frac{C_L}{\dot{m}_1} \quad (1)$$

where:

UCL = unit cost of labor ($/Mg);

C_L = cost of labor to operate the plant ($/hr); and

$\dot{m}_1$ = mass flowrate of raw MSW.

The hourly cost of labor (C_L) is equal to the product of the wage rate and the number of machine operators and spotters, i.e.,

$$C_L = W_m \sum_{k=1}^{p} N_{mk} + W_s \sum_{k=1}^{p} N_{sk} \quad (2)$$

where:

W_m = fully burdened wage rate for equipment operators;
W_s = fully burdened wage rate for spotters;
N_m = number of equipment operators in plant;
N_s = number of spotters in plant;
k = group of unit operations; and
p = number of unit operation groups in plant.

Equipment operators are those who control machinery from a control room and monitor operations via closed circuit television. Spotters monitor, and sometimes control, processes in situ and have verbal contact with the equipment operator(s) via radio or telephone.

The unit operation groups (UOG's) comprise one or more unit operations, and each operation includes one or more pieces of equipment. The number of pieces of equipment in each unit operation is selected by the designer. The UOG's are as follows:

UOG 1 (k=1) comprises unit operations other than conveying that receive raw MSW.

UOG 2 (k=2) comprises any one of the following unit operations that immediately follows UOG 1: size reduction, screening, and air classification.

UOG 3 (k=3) includes any two of the three unit operations specified under UOG 2 that are not included in UOG 1 or UOG 2. It should be noted that a unit operation may occur at more than one place in the processing system and is considered a separate unit operation each time it occurs.

UOG 4,5,6,etc. (k=4,5,6,etc.) include three of the unit operations specified under UOG 2 that are not included in a previous UOG.

UOG p (k=p) includes only densification.

Ferrous material separation and conveying are not included in any unit operation group because, in typical operations, they do not warrant the utilization of a significant amount of marginal operating labor.

Figure 3.1 illustrates the grouping of unit operations into unit operation groups in a hypothetical processing plant. In some instances, the assignment of a unit operation to a particular UOG is arbitrary. For example, air classification could have been assigned to UOG 3 rather than to UOG 4; and secondary size reduction or fine screening could have been included in UOG 4 rather than in UOG 3.

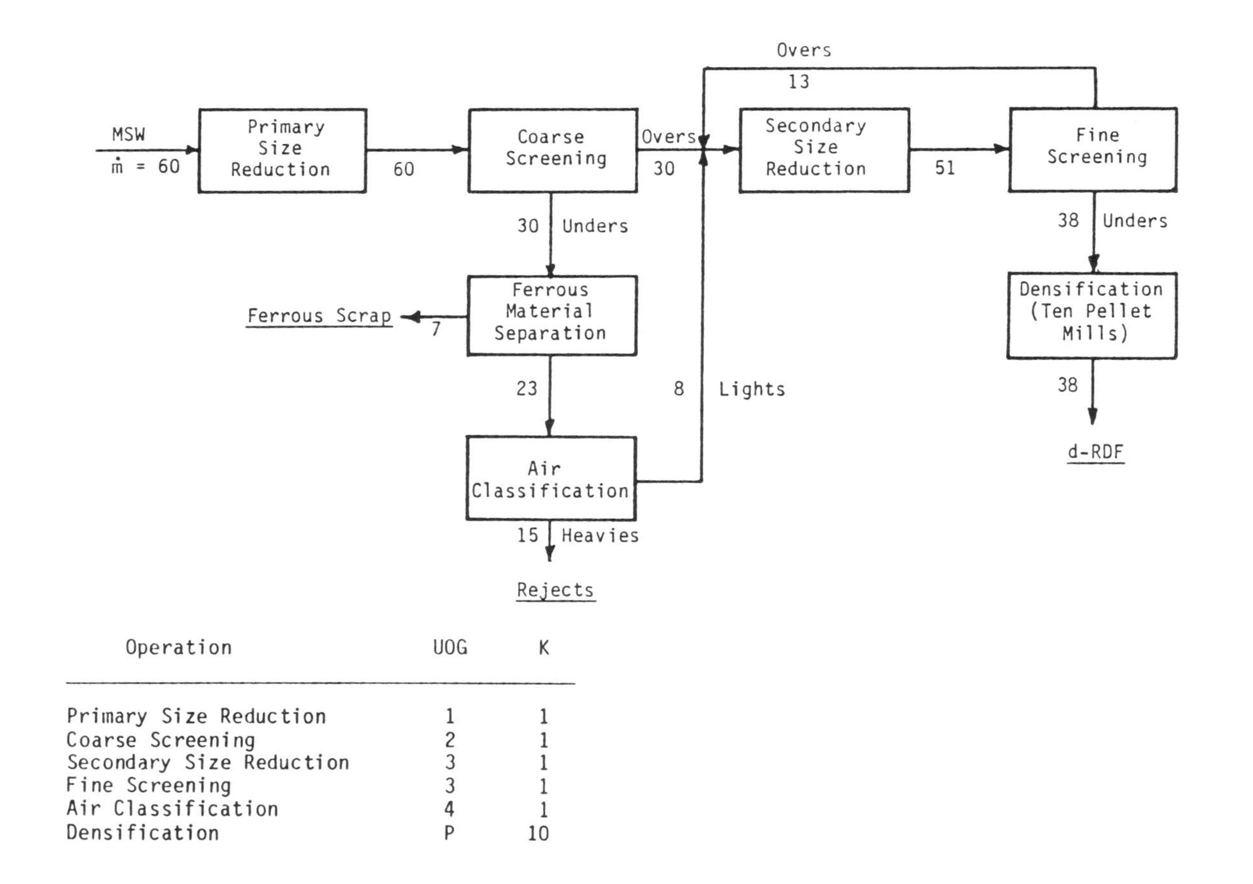

Operation	UOG	K
Primary Size Reduction	1	1
Coarse Screening	2	1
Secondary Size Reduction	3	1
Fine Screening	3	1
Air Classification	4	1
Densification	P	10

Figure 3.1. Illustration of Nomenclature Used in Determining Labor Requirements

In the operating cost model, the wage rates (W_m and W_s in Eq. 2) are inputs to the model as are the number of UOG's (p).

For the three unit operations, size reduction, screening, and air classification, the operating labor requirement is calculated by first computing a 'use factor' viz

$$A_k = \sum_{i=1}^{n_k} \frac{\dot{m}_{ki}}{\dot{m}_{ki}+10} \tag{3}$$

where:

A_k = use factor for UOG k;

$\dot{m}_{ki}$ = mass throughput rate (Mg/hr) for processing line i of UOG k; and

n_k = number of processing lines in UOG k.

For the system shown in Figure 3.1, the use factor for unit operation group 3 is calculated as follows (note that n_k equals two because there are two processing lines, one for secondary size reduction and one for fine screening):

$$A_3 = \sum_{i=1}^{2} \frac{\dot{m}_{3i}}{\dot{m}_{3i}+10} \tag{4}$$

$$= \frac{51}{51+10} + \frac{51}{51+10}$$

$$= 1.67$$

By similar calculations, the values of A_1, A_2, and A_4 are determined,

A_1 = 0.86

A_2 = 0.86

A_4 = 0.70

Except when $A_1 + A_2 < 1$, the number of machine operators and spotters is determined by

$$N_{Tk} = \text{INT}\left(\frac{A_k}{2} + 1\right) \tag{5}$$

where:

N_{Tk} = number of machine operators plus number of spotters required for UOG k.

INT indicates that only the integer part of the quantity in brackets is taken. That is, the value in brackets is rounded downward to a whole number.

If $A_1 + A_2 < 1$, then N_{T2} equals zero and N_{T1} is calculated from Eq. 5.

Letting

N_{mk} = number of machine operators required for UOG k, and

N_{sk} = number of spotters required for UOG k

and noting that

$$N_{Tk} = N_{mk} + N_{sk}$$

the number of machine operators and spotters is given by

$$N_{Tk} = N_{mk} \text{ when } k=1 \tag{6a}$$

$$N_{mk} = \text{INT}\left(\frac{N_{Tk}}{2}\right) \text{ when } k=2 \text{ or } p>k>3 \tag{6b}$$

$$N_{sk} = \text{INT}\left(\frac{N_{Tk}}{2}\right) \text{ when } k=3 \tag{6c}$$

Using Eqs. 5 and 6 to determine the labor requirements for the system shown in Figure 3.1 (excluding densification) yields,

$$N_{T1} = \text{INT}\left(\frac{0.86}{2} + 1\right) = \text{INT}\ (1.43) = 1$$

$$N_{T2} = 1$$

$$N_{T3} = 1$$

$$N_{T4} = 1$$

$$N_{m1} = 1;\ N_{s1} = 0$$

$$N_{m2} = 0;\ N_{s2} = 1$$

$$N_{m3} = 1;\ N_{s3} = 0$$

$$N_{m4} = 0;\ N_{s4} = 1$$

Thus, two equipment operators and two spotters are required.

The labor requirement for densification (k=p) is founded on the estimation that one spotter can operate up to four pellet mills. Thus,

$$N_{Tp} = N_{sp}$$

$$\text{If INT}\left(\frac{n_p}{4}\right) = \frac{n_p}{4}, \text{ then } N_{sp} = \frac{n_p}{4} \tag{7a}$$

$$\text{If INT}\left(\frac{n_p}{4}\right) < \frac{n_p}{4}, \text{ then } N_{sp} = \text{INT}\left(\frac{n_p}{4}\right) + 1 \tag{7b}$$

In Figure 3.1, there are ten pellet mills ($n_p = 10$). Then

$$\text{INT}\left(\frac{10}{4}\right) < \frac{10}{4}$$

So

$$N_{sp} = \text{INT } (2.5) + 1$$

$$= 3$$

That is, three pellet mill operators are required.

The entire system shown in Figure 3.1, then, requires two equipment operators in the control room and five spotters. Assuming a fully burdened wage rate of $25/hr for both equipment operators and spotters yields, from Eq. 2,

$$C_L = 25(2) + 25(5)$$
$$= \$175/\text{hr}$$

Substituting this value in Eq. 1 and noting that the mass input to the system is 60 Mg/hr yields a unit cost for labor of $2.92/Mg.

3.2 ENERGY

The specific energy requirements (i.e., energy per unit mass of feedstock to a given unit process) for each unit process (E_o) are presented in the Task 3 Report and are summarized in Table 3.1. All of the energy is in the form of electricity.

The unit cost of energy for unit operation j is

$$UCE_j = PE_{oj} \tag{8}$$

Table 3.1. Specific Energy Requirements

Unit Operation	Equipment	Specific Energy (kWh/Mg)
Size Reduction	Hammermill Shredder	$\left(\sum_{i=1}^{n} \dot{m}_i A_i Z_{oi}^{b_i} \right) \Big/ \sum_{i=1}^{n} \dot{m}_i$
Air Classification	Vertical Air Classifier with Baghouse	$(9.63 + 0.001 v^2) Q/\dot{m}$
Screening	Trommel Screen	0.4
Ferrous Separation	Multiple Stage Cross-Belt Magnet	0.36
Densification	Rotary-Die Pellet Mill	See Figure 5.3 in Task 3 Report
Conveying	Belt Conveyor	$2.1 \left(\frac{1}{\rho\dot{m}}\right)^{1/2} (1 + 0.021 L) + 0.0044 (1 + 0.021 L + 0.69 H)$
	Apron Conveyor	$(L^2 - H^2)^{1/2} (0.11 W v/\dot{m} + 0.0003) + 0.003 H$

Note: The equations given in this table are described in the Task 3 Report.

where:

UCE_j = unit cost of energy for unit operation j ($/Mg);
P = price of electrical energy ($/kWh); and
E_{oj} = specific energy requirement for unit operation j.

The price is an input to the model.

The unit cost of energy for the entire processing system is given on the basis of a unit mass of MSW feedstock to the plant,

$$UCE_s = \frac{\sum_j (UCE_j \, \dot{m}_j)}{\dot{m}_s} \tag{9}$$

where:

UCE_s = unit cost of energy for the processing system ($/Mg);
$\dot{m}_j$ = mass throughput rate to unit process j (Mg/hr); and
$\dot{m}_s$ = mass throughput rate to the processing system (Mg/hr).

4. Maintenance Costs

The unit cost of maintenance for each unit operation is given in the following section. The costs are summarized in Table 4.1.

4.1 HAMMERMILL SHREDDER

In studying the operation of size reduction equipment in several MSW processing systems, CRS [1] found maintenance labor costs for primary shredding to range from $0.003/Mg to $1.12/Mg, with an average of $0.26/Mg (the figures in the report are increased by 50 percent to account for the $30/hr wage rate used in this model). The cost of hardfacing alloy and hammer replacement is reported by CRS as $0.18/Mg and $0.35/Mg respectively for primary shredding. The total maintenance cost is thus, $0.79/Mg if no extraordinary repairs are required. For secondary shredding, the CRS report cites a labor cost of $0.48/Mg and hammer replacement costs of $0.23/Mg. The costs of hardfacing alloy were not determined in the CRS study.

Table 4.1. Unit Costs of Maintenance

Unit Operation	Type of Equipment	Unit Cost ($/Mg)
Size Reduction (Primary and Secondary)	Hammermill Shredder	0.80
Air Classification	Vertical Air Classifier with Cyclone and Baghouse	0.30
Screening	Trommel Screen	0.20
Ferrous Material Recovery	Cross belt Magnet	0.02
Densification	Pellet Mill	
SLF Feedstock		3.70
ACLF Feedstock		5.00
Conveying	Belt and Apron Conveyors	$1.5 \times 10^{-5} \times \frac{C_c}{\dot{m}}$

A review of maintenance costs for the Ames resource recovery plant in 1983 indicates a unit maintenance cost of $0.82/Mg, expressed in 1984 dollars.

A unit cost of $0.80/Mg is used in this model for both primary and secondary size reduction via a hammermill shredder.

4.2 AIR CLASSIFICATION

In studying seven air classifiers at MSW processing plants, Midwest Research Institute (MRI) was able to obtain reliable maintenance data only from the plant at Ames, Iowa [2]. Based upon the MRI report, the cost of maintenance and electricity is $0.75/Mg when expressed in 1984 dollars. Electricity consumption was 7.5 kWh/Mg, and cost $.057/kWh (1984). This leaves about $0.30/Mg for maintenance which is the figure used in this model.

4.3 TROMMEL SCREENING

In a detailed study of the trommel screen at the Baltimore County plant [3], Midwest Research Institute found the maintenance costs of the 10 Mg/hr screen to be about $0.30/Mg (assuming $30/hr for labor). Inasmuch as the screen suffered breakdowns, such as the repeated cracking of a carrier ring, that would not be expected in an optimally designed and operated screen, the maintenance cost for trommel screening is taken to be $0.20/Mg.

4.4 FERROUS MATERIAL RECOVERY

Manufacturers indicated that the belt is the most likely source of maintenance costs. The belt is generally covered with metal pads that must be replaced when worn or displaced. The pads typically cost less than $100 each, and at the Ames RDF plant only a few need replacing each year. Since the plant processes more than 60,000 Mg of MSW per year, the cost of pads amounts to less than $0.01/Mg. In the model, a unit maintenance cost of $0.02/Mg is used. This would correspond to one hour of labor per week and $400 of supplies per year in a magnetic separation system receiving 50 Mg of feedstock per hour (100,000 Mg/yr). The annual cost of labor and supplies is assumed to vary linearly with the throughput.

4.5 DENSIFICATION

Information from two manufacturers of pellet mills regarding maintenance costs can be summarized as follows:

1. The total cost of maintenance is about twice the cost of die maintenance or about 170 percent of the cost of die and roller maintenance.
2. Dies have a useful life of 3600 to 6400 Mg when screened light fraction (SLF) with an ash content of about 10 percent is the feedstock. The die life is 3400 to 4800 Mg when air classified light fraction (ACLF) of 20 to 25 percent ash content is the feedstock.
3. Dies must be reconditioned once during their life at a cost of about $1500 each.
4. Up to four rollers are replaced for each die replacement. Rollers cost about $900 each.

From these data, the expected total unit maintenance cost for densification of SLF and ACLF is estimated to be $3.7/Mg and $5.0/Mg respectively.

4.6 CONVEYORS

The cost of maintaining conveyors is dependent on several factors such as the abrasiveness of the material being conveyed, the height from which the material is dropped onto the conveyor, etc. Since there is a paucity of data on the cost of maintaining conveyors for various material fractions in MSW processing plants, a cost of 3 percent of the installed capital cost per 2000 operating hours is used in this model. The unit cost of maintenance is

$$UCM_c = 1.5 \times 10^{-5} \frac{C_c}{\dot{m}_c} \qquad (1)$$

where:

UCM_c = unit cost of maintenance of conveyors ($/Mg);
C_c = installed cost of conveyors ($); and
$\dot{m}_c$ = mass flowrate on conveyor (Mg/hr).

As a point of reference, it may be noted that, according to this equation, a horizontal belt conveyor carrying 15 Mg/hr of material with a density of 100 kg/m^3 and being 20 m long and 1 m wide would cost $0.05/Mg or $1580/yr to maintain.

4.7 REFERENCES

1. Savage, G.M., D.J. Lafrenz, D.B. Jones, and J.C. Glaub, Engineering Design Manual for Solid Waste Reduction Equipment, prepared for the U.S. EPA under Contract No. 68-03-2972, 1982.

2. V. Hopkins, B.W. Simister, and G.M. Savage, Comparative Study of Air Classifiers, prepared for the U.S. EPA under Contract No. 68-03-2730, 1981.

3. Hennon, G.J., D.E. Fiscus, J.C. Glaub, and G.M. Savage, An Economic and Engineering Analysis of a Selected Full-Scale Trommel Screen Operation, prepared for the Department of Energy under Contract No. DE-AC03-80CS24330, Oct. 1983.

5. Residue Disposal Costs

The residue disposal costs are modeled on a plant-wide basis rather than on a unit process basis, because the output from one unit operation is often the feedstock to another unit operation. The unit cost of residue disposal is,

$$UCR = mf_r \times C_r \tag{1}$$

where:

UCR = unit cost of residue disposal ($/Mg of feedstock to the system);

mf_r = mass fraction of MSW feedstock that is disposed; and

C_r = cost of disposing residue ($/Mg).

The cost of disposal, C_r, includes the cost of transporting the residue from the processing plant to the disposal site as well as the cost of disposal. The mass fraction to be disposed (mf_r) is an input to the economic model and is determined from the mass balance model. It is equal to one minus the sum of the mass fractions recovered as salable materials.

6. Revenue

The income from the sale of recovered materials is modeled on a plant-wide basis, because allocating revenues to a particular unit operation is arbitrary when more than one unit operation contributes to the recovery and upgrading of the material. The unit revenue is

$$UR = (mf_{fe} \times P_{fe}) + (mf_{rdf} \times P_{rdf}) \quad (1)$$

where:

UR = unit revenue from the sale of recovered materials ($/Mg of raw MSW feedstock);

mf_{fe} = mass fraction of raw MSW recovered as ferrous scrap;

mf_{rdf} = mass fraction of raw MSW recovered as refuse derived fuel;

P_{fe} = price of ferrous scrap, FOB the processing plant; and

P_{rdf} = price of refuse derived fuel, FOB the processing plant.

The mass fractions are determined from the mass balance models. The prices are inputs to the model and are assigned by the user of the model.

Part IV

Task 5 Report: Unit Operation Models: Mass Balance, Energy Requirements, and Economics

Preface

The Task 5 Report presents the models for the mass balance, energy requirements, and economics of the following generic unit operations:

- Shear Shredding
- Disc Screening
- Glass Separation
- Non-Ferrous Separation

Mass balance and energy requirement models are presented in Sections 1 through 4; and the economic model is presented in Section 5.

Also presented in the report are the derivations of the models and their limitations.

1. Shear Shredder Model

1.1 BACKGROUND

The shear shredder is receiving attention as a means of achieving the desired degree of size reduction of the combustible portion of MSW, while avoiding the pulverization of the inert material, such as glass. Shear-shredded solid waste, consequently, has been purported to be more suitable for removal of inerts for production of high quality RDF than the product resulting from high-speed shredding, such as hammermilling.

Shear shredders, unlike hammermills, are by design low-speed size reduction devices. They accomplish size reduction through tearing and shearing effected by offset counter-rotating cutters. The size reduced material falls through the spaces among the cutters. An additional purported, although undocumented, advantage of shear shredders is that their low-speed design results in a lower potential for explosions than high-speed size reduction units.

The size reduction model for shear shredding is formulated to describe the product size distribution and energy requirements associated with solid waste comminution under given conditions of feed size distribution and of machine configuration. The model is structured to simulate refuse comminution as a function of the type of components comprising the feedstock. No detailed published literature is available on the size reduction performance of shear shredders. Consequently, the development of the shear shredder model follows entirely from a theoretical approach.

The basic block diagram for the shear shredder model is shown in Figure 1.1.

1.2 DEVELOPMENT OF MASS BALANCE MODEL

1.2.1 Description of Model

The solid waste size reduction model has been developed using the concept of linear, size discrete comminution kinetics. The approach has been followed previously by a number of researchers in the field of

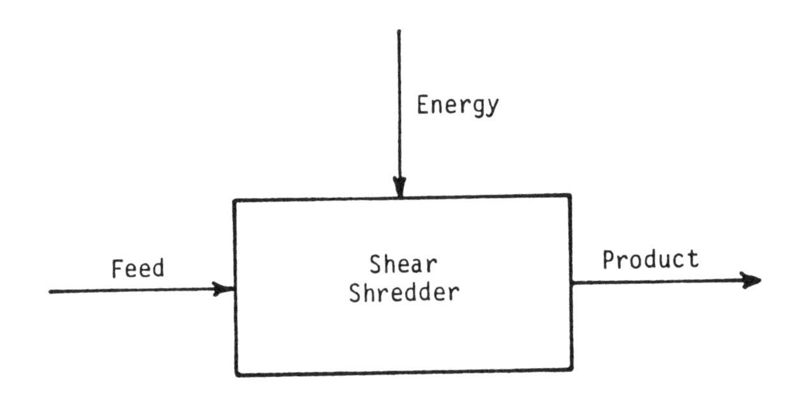

Figure 1.1. Block Diagram of the Shear Shredder Model

mineral comminution. With regard to the development of a model for the shear shredding of solid waste, the linear, size discrete approach allows convenient matrix representation of the process. Thus, the representation is amenable to simulation by digital computer. In the development of the size reduction model, the governing equations of comminution are derived initially from the consideration of the batch milling process and are subsequently extended to continuous steady-state milling by invoking the concept of the residence time distribution of material within the mill cavity.

The governing relations are developed for a single-component material and extended to multi-component size reduction through the assumption of linearity, i.e., the breakage behavior of each component is considered to be independent of the presence of other components. The model computes the size distribution for each component. The cumulative size distribution for the mixture is computed as the sum of the products of the component size distributions and their respective mass fractions present in the feed.

The model uses the concept of selection and breakage functions to describe the breakage of material within the size reduction device. The use of the functions allows different types of size reduction devices to be modeled inasmuch as both selection and breakage events are governed to a large degree by the internal geometry of the mill and by its operating conditions. In addition to the machine parameters, the properties of the material also influence its breakage within the mill.

1.2.2 Important Assumptions

The important assumptions used in formulating the size reduction model are recapped below:

1. The throughput is constant. Therefore, size reduction is accomplished under steady-state conditions.
2. The breakage behavior of each component is independent of the presence of other components.
3. S_1 and B_{ij} are independent of size class and time. Therefore, the cumulative breakage function can be normalized for each size class of the feed, and the kinetic model is linear with constant coefficients.
4. The values of the breakage function are represented by the relation,

$$B_{ij} = \frac{K_1}{S_1}\left(\frac{X_i^2}{X_j X_{j+1}}\right)^{\alpha/2}$$

where:

K_1 and α are constants;

S_1 = the selection function value for the top size class; and

X = mass fraction in a given size class.

5. Residence time (τ) is related to mass throughput (Q) according to an equation of the form,

   $\tau = aQ^b$

6. All size classes for a given component have identical residence time distributions.

1.2.3 Governing Theory

The linear, size-discrete matrix model for size reduction is formulated by dividing the feed into discrete narrow size classes. Establishing a mass balance on the material in each size interval results in the following relation [4] for open-circuit batch milling,

$$d\,\frac{(H\ m_i(t))}{dt} = -\ S_i\ H\ m_i(t) + \sum_{j=1}^{i-1} b_{ij}\ S_j\ H\ m_j(t) \tag{1}$$

where:

$m_i(t)$ = mass fraction in the ith size class;

H = total mass of material in the size reduction device at time t;

S_i = fractional rate at which material is broken out of the ith size class; and

b_{ij} = fraction of material in the jth size class that appears in the ith size class.

Invoking the assumption that S_i and b_{ij} are independent of size class and time, Eq. 1 may be rewritten in matrix notation for the complete ensemble of size classes,

$$d\,\frac{[H\ \underline{m}(t)]}{dt} = -\ [\underline{\underline{I}}-\underline{\underline{B}}]\ \underline{\underline{S}}\ H\ \underline{m}(t) \tag{2}$$

where:

$\underline{\underline{I}}$ = identity matrix;

$\underline{\underline{S}}$ = selection matrix; and

$\underline{\underline{B}}$ = breakage function matrix.

In steady-state operation, the mass of material in the size reduction device is constant. Thus, for steady-state conditions Eq. 2 becomes,

$$d\frac{[\underline{m}(t)]}{dt} = -[\underline{\underline{I}}-\underline{\underline{B}}]\,\underline{\underline{S}}\,\underline{m}(t) \tag{3}$$

for which the analytical solution is,

$$\underline{m}_b(t) = \exp\,[-(\underline{\underline{I}}-\underline{\underline{B}})\,\underline{\underline{S}}\,t]\,m_b(o) \tag{4}$$

where:

$m_b(o)$ = initial mass of material in the mill.

The subscript b denotes that the solution is for batch comminution.

In the case where no two selection functions are equal, the term exp $[-(\underline{\underline{I}}-\underline{\underline{B}})\,\underline{\underline{S}}\,t]$ can be simplified by a similarity transform [5]. Thus, Eq. 4 is transformed to give,

$$\underline{m}_b(t) = \underline{\underline{T}}\,\underline{\underline{J}}(t)\,\underline{\underline{T}}^{-1}\,m_b(o) \tag{5}$$

where:

$$T_{ij} = \begin{cases} 0 & i<j \\ 1 & i=j \\ \sum_{k=j}^{i-1} \dfrac{b_{ik}\,S_k}{S_i - S_j}\,T_{kj} & i>j \end{cases}$$

$$J_{ij}(t) = \begin{cases} \exp\,(-S_i t) & i=j \\ 0 & i\neq j \end{cases}$$

Inasmuch as a continuous size reduction relation is sought for the refuse comminution model to be developed here, the batch comminution relation (Eq. 5) must be extended to continuous milling. The extension is made using the concept of residence time distribution [3]. The residence time distribution is defined as the mass of material of a given size that is contained within the size reduction device as a function of time. If it is assumed that a single residence time distribution characterizes all of the particle size classes, the steady-state size distribution from the mill can be represented by an average of the batch responses weighted with respect to the residence time distribution, i.e.,

$$m_{cp} = \int_0^\infty m_b(t)\ R(t)\ dt \qquad (6)$$

where:

$R(t)$ = residence time distribution; and

cp = product under continuous milling conditions.

Substituting the transformed relation for $m_b(t)$ (i.e., Eq. 5) into Eq. 6 yields,

$$\underline{m}_{cp} = \underline{\underline{T}}\ [\int_0^\infty \underline{\underline{J}}(t)\ R(t)\ dt]\ \underline{\underline{T}}^{-1}\ \underline{m}_{cf} \qquad (7)$$

where:

m_{cf} = mass fraction of the feed material in continuous mill operation.

The integrand in Eq. 7 is commonly expressed in terms of the dimensionless time variable, $\theta = t/\tau$, where τ is the mean resident time and is the quotient of the mass of material held within the mill and the throughput. The mean residence time τ is assumed to be related to the throughput (Q) through an equation of the form [6],

$$\tau = aQ^b \qquad (8)$$

where:

a and b are constants that represent the characteristics of the mill.

Expressing Eq. 7 in terms of θ yields,

$$\underline{m}_{mp} = \underline{\underline{T}}\ \underline{\underline{J}}_c(\tau)\ \underline{\underline{T}}^{-1}\ \underline{m}_{mf} \qquad (9)$$

where:

$$J_{c_{ij}}(\tau) = \begin{cases} \int_0^\infty R(\theta)\ \exp\ (-S_i\tau\theta)\ d\theta & i=j \\ 0 & i \neq j \end{cases}$$

To model refuse size reduction devices, the continuous open circuit relation given in Eq. 9 has been modified by the addition of an internal classifier function [4]. The classifier represents the openings (or clearance dimension) that restrict the flow of material through the mill until the size of the particles is less than the size of the openings. In the case of the shear shredder, the classifier function simulates the spacing among the counter rotating cutters. The steady-state description of the refuse size reduction device follows from Eq. 9 and a mass balance on the size reduction equipment [6]. A representation of the size reduction circuit is shown in Figure 1.2. Since under steady-state conditions the infeed and discharging rates are identical, the governing relation for closed circuit size reduction is

$$\underline{m}_p = [\underline{\underline{I}}-\underline{\underline{C}}]\ \underline{\underline{I}}\ \underline{\underline{J}}_c(\tau)\ \underline{\underline{T}}^{-1}\ [\underline{\underline{I}}-\underline{\underline{C}}\ \underline{\underline{T}}\ \underline{\underline{J}}_c(\tau)\ \underline{\underline{T}}^{-1}]^{-1}\ \underline{m}_f \qquad (10)$$

The size discrete selection function describes the mass fraction within a discrete size class that is selected for breakage. In the refuse size reduction model a value (S_1) is chosen for the mass fraction selected for breakage in the top size class. The values of the selection function for the smaller size classes are computed from the cumulative breakage function.

The estimation of the breakage function follows from two assumptions. First, the size reduction process is linear, i.e., the values of the breakage function are independent of the size distribution of material in the mill. Secondly, the size discrete breakage function can be normalized, i.e., for material breaking into smaller size classes there is a constant ratio of breakage values that is dependent upon the ratio of the successive size intervals.

The size reduction model uses a breakage function relation presented by Epstein [7] for mineral comminution and used by Shiflett [3,6] to model refuse size reduction in a horizontal swing hammermill,

$$B_{ij} = \frac{K_1}{S_1}\left(\frac{X_i^{\,2}}{X_j\,X_{j+1}}\right)^{\alpha/2} \qquad (11)$$

where:

X = size class; and

K_1 = an invariant constant.

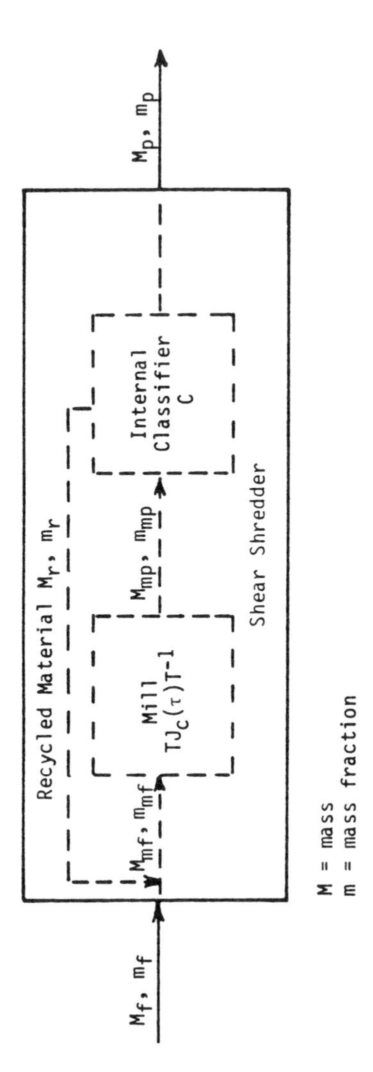

Figure 1.2. Diagram of a Shear Shredder with an Internal Classifier

Inasmuch as the cumulative breakage function value for the top size class is constrained by the mass balance to a value of unity, the value $K_1 [X_i/(X_j X_{j+1})]^{\alpha/2}$ has been set equal to S_1 in the size reduction model for each of the top size classes. The imposition of the above constraint is a departure from the constraints imposed by Shiflett [3,6]. The constraint, however, is a necessary one in order to uphold the conservation of mass.

The computation of the cumulative breakage function values requires that α be specified. For the present model α is chosen, in addition to the S_1 value, to give an empirical fit between a set of measured product size distributions and the set of predicted values. In the case of MSW, the goodness-of-fit is constrained by the form of Eq. 11.

Values of S_1 and α are typically determined from field test data. The definition of values for S_1 and α for shear shredding must await the development and analysis of the appropriate field test data.

The following is a listing of the key inputs to the size reduction model:

- Number of Components
- Number of Size Classes
- Number of Residence Time Intervals
- MSW Composition
- Size Class Designations
- Classifier Function Values
- Residence Time/Throughput Equation Constants by Component
- Alpha (α) Values by Component
- S_1 Values by Component
- Residence Time Distribution by Component
- Component Feed Size Cumulative Percent Passing Values
- Throughput

A detailed flow chart of the size reduction model is shown in Figure 1.3.

1.3 ENERGY REQUIREMENTS

The energy model for shear shredding is characterized in terms of components, similar to the development of the mass balance model. The

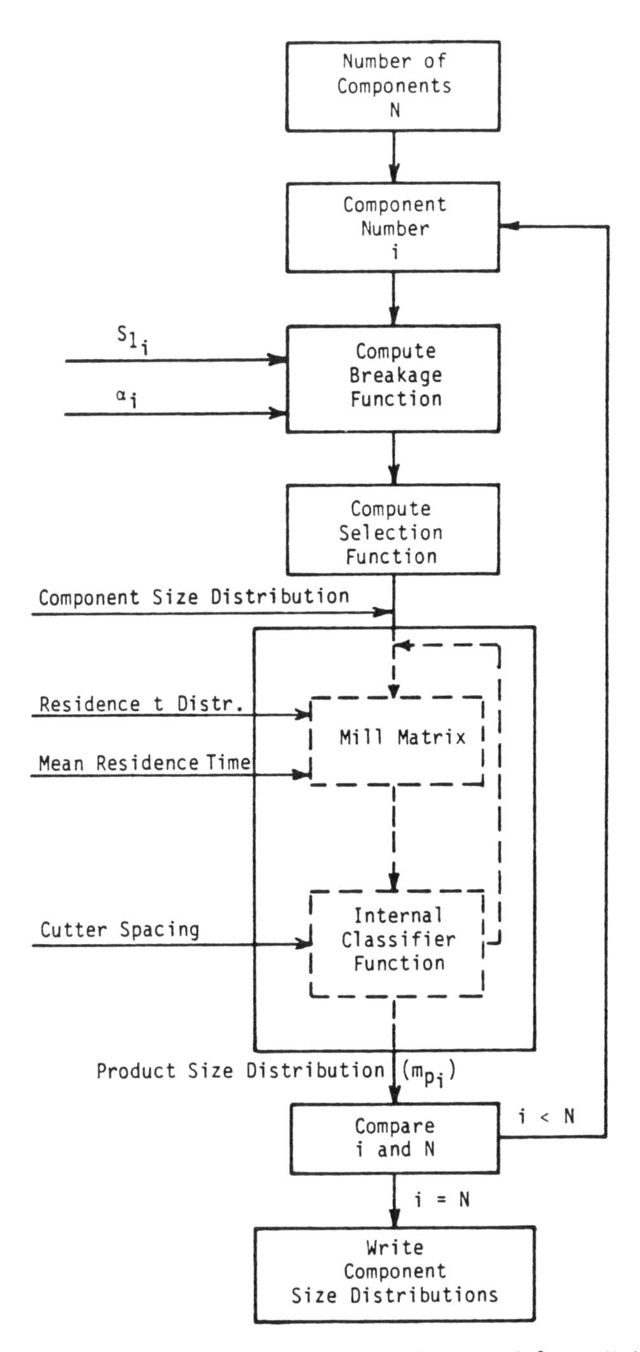

Figure 1.3. Flow Chart of Shear Shredder Mass Balance Model

approach follows that employed for the energy model presented previously for hammermills in the Task 3 Report.

The governing relation for the energy requirements of shear shredding is cast in terms of the specific energy (E_0, kWh/Mg) required to achieve a given degree of size reduction (Z). The parameter Z is expressed in terms of the characteristic size of the feed (F_0) and of the product (P_0). The values of F_0 and P_0, respectively, are numerically interpolated from the input feed size data and the product size distribution calculated by the size reduction model. The general form of the energy equation is,

$$E_0 = A\ Z^B \tag{12}$$

where:

A and B are empirically determined coefficients and $Z = (F_0 - P_0)/F_0$.

(The characteristic size is that size corresponding to 63.2 percent cumulative passing.)

Lacking the appropriate experimental data, the coefficients A and B for the various waste components cannot be determined for shear shredding.

No comparison of the predicted and actual energy requirements for shear shredding is possible due to a lack of appropriate experimental data.

1.4 REFERENCES

1. Obeng, D.M., Comminution of a Heterogeneous Mixture of Brittle and Non-brittle Materials, Ph.D. thesis, University of California, Berkeley, 1974.

2. Obeng, D.M., and G.J. Trezek, "Simulation of the Comminution of a Heterogeneous Mixture of Brittle and Non-brittle Materials in a Swing Hammermill," Ind. Eng. Chem. Process Des. Dev., 14, No. 2, 1975.

3. Shiflett, G.R., and G.J. Trezek, "The Use of Residence Time and Nonlinear Optimization in Predicting Comminution Parameters in the Swing Hammermilling of Refuse," Ind. Eng. Chem. Process Des. Dev., Vol. 18, No. 3, 1979.

4. Herbst, J.A., G.A. Grandy, and D.W. Fuerstenau, "Population Balance Models for the Deisgn of Continuous Grinding Mills," Proc. of the Tenth IMPC, London, 1973.

5. Pease, M.C., Methods of Matrix Algebra, Academic Press, New York, 1965.

6. Shiflett, G.R., "A Model for the Swing-Hammermill Size Reduction of Residential Refuse," D. Eng. dissertation, University of California, Berkeley, 1978.

7. Epstein, B., "The Mathematical Description of Certain Breakage Mechanisms Leading to the Logarithmico-Normal Distribution, "J. Frank. Inst., 224, 1947.

2. Disc Screening Model

2.1 BACKGROUND

Disc screens consist of a series of rotary shafts onto which a number of parallel discs are attached. The discs from each alternating shaft are intermeshed. As the shafts rotate, particles smaller than the openings presented by the disc faces and shafts fall through the openings. Larger particles are transported along the top of the screen to the discharge end by the action of the rotating discs.

Disc screens are used in refuse processing for removing inorganic materials from a fuel fraction and for particle size control. A simplified block diagram of a disc screening process is shown in Figure 2.1. The material that passes through the screen is commonly referred to as either the "undersize" or the "unders." The material that does not pass through the screen is commonly referred to as the "oversize" or the "overs." However, a distinction must be made between the true oversize and the material that does not pass through the screen, inasmuch as some undersize material is present in the material that exits the screen at the downstream end. For the purposes of the development of the disc screen model, the terms "unders" and "overs" are used to describe the split streams leaving the screen.

In all of the resource recovery facilities where disc screens have been installed, the wastes are shredded prior to the screening operation. Various other unit processes, such as magnetic separation and air classification, also have been employed prior to the disc screen. Typically, disc screens are employed to remove inert materials from shredded MSW and to scalp oversize material for size reduction. The screens are predominantly used to improve the quality of RDF.

The key parameters affecting the performance of a disc screening operation can be categorized as construction parameters, operating parameters, and feed characteristics. In general, the only operating parameters over which control can be exerted are the feed rate to the screen and the rotational speed of the shafts. However, modification of

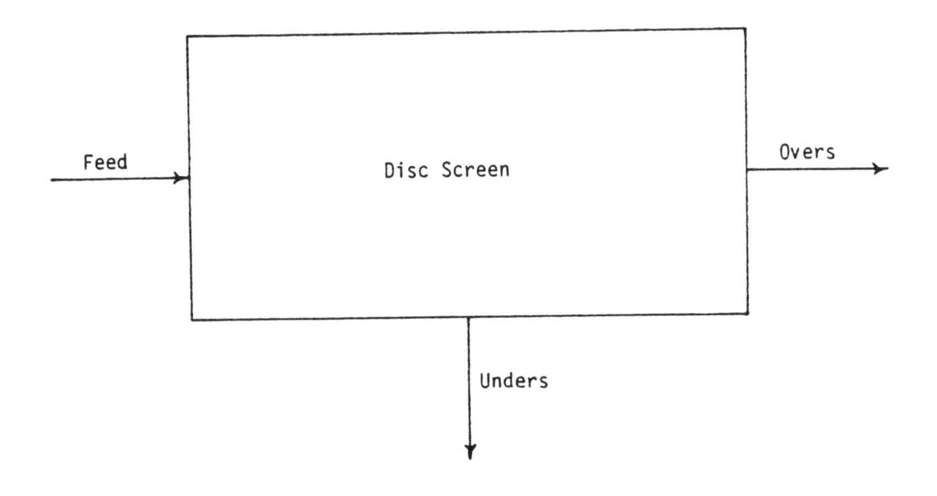

Figure 2.1. Block Diagram of Disc Screening Process

aperture size may be somewhat easier than that for trommel screens inasmuch as aperture size can be changed by replacement of shafts with shafts having different disc shapes and spacings.

The parameter most commonly used for the characterization of screening performance is the screening efficiency. This parameter represents the percentage of undersize material entering the screen that passes through the apertures in the screen to the undersize conveyor belt. In equation form, screening efficiency (η) can be formulated in several ways. The following form is employed in the development of the disc screening model:

$$\eta = \frac{\dot{m}_u}{\dot{m}_u + U_o \dot{m}_o} \tag{1}$$

where:

$\dot{m}_u$ = flowrate of the undersize fraction;
$\dot{m}_o$ = flowrate of the oversize fraction; and
U_o = fraction of true undersize material in the oversize fraction.

2.2 DEVELOPMENT OF MASS BALANCE MODEL

2.2.1 Description of Model

To the knowledge of the authors, the published literature does not contain any models or analytical relations for the disc screening process. Moreover, very little field test information has been collected and presented in the literature. Thus, the model presentation that follows is the first published attempt towards simulating the disc screening of MSW. The modeling efforts have been complicated and restricted by the lack of thorough field test data.

The basic structure of the disc screening model is similar to that of the trommel screening model presented previously. However, different relations are developed for the two primary aspects of particle behavior in the screening process, namely: (1) the particle dynamics, which determine the number of collisions between particles and the screen surface; and (2) the probability of passage when a particle strikes the surface. A diagrammatic illustration of the model is shown in Figure 2.2.

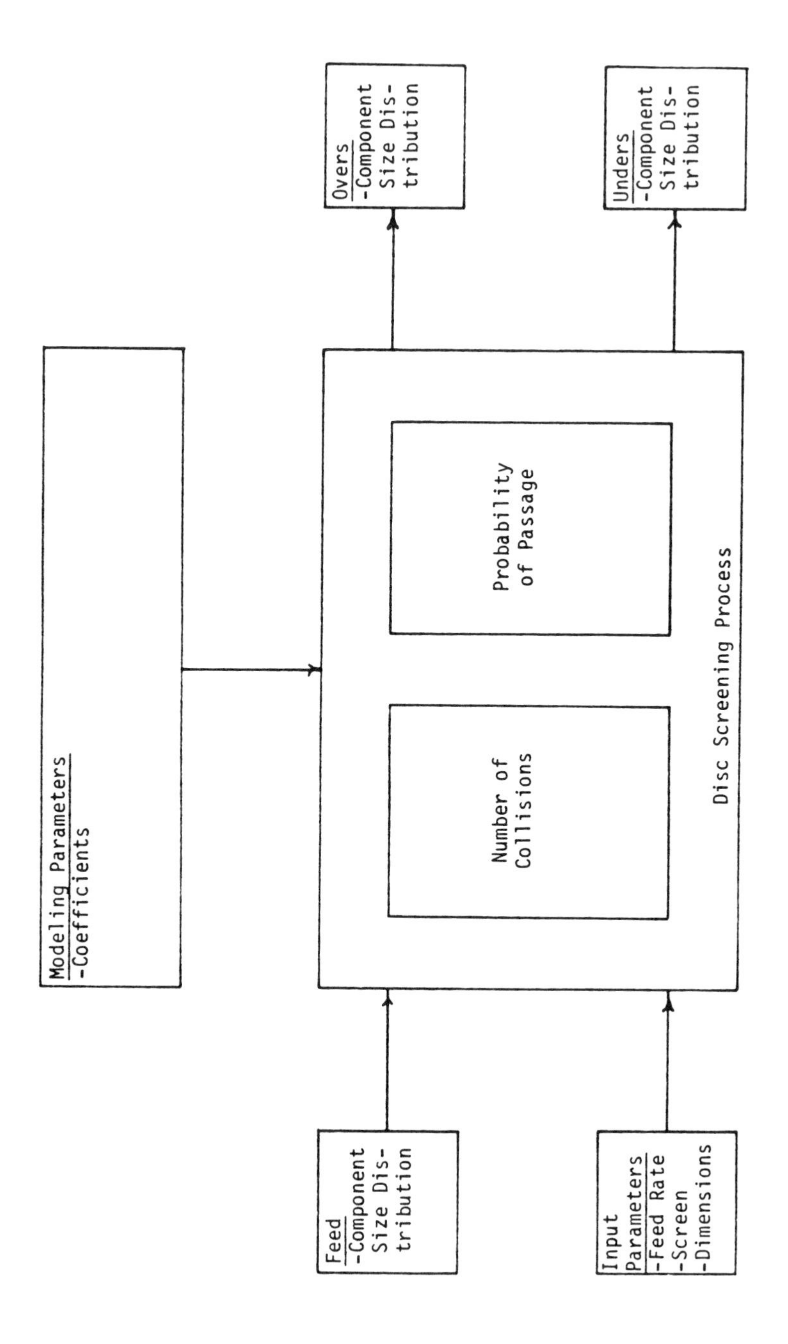

Figure 2.2 Block Diagram of Disc Screening Model

The model is structured such that a component size feed matrix is input to the model and the model predicts the component size matrices of the overs stream and of the unders stream.

Unlike the trommel screening process, the number of apertures which a particle encounters in a disc screen is fairly well defined. The number of apertures over which a particle travels is essentially equal to the number of shafts minus one. This statement assumes that the material entering the screen falls onto the first shaft or the first opening between shafts. As a first approximation then, the number of apertures which a particle encounters may be given by

$$N_c = N_s - 1 \qquad (2)$$

where:

N_c = calculated number of contacts; and

N_s = number of shafts.

To account for the reduced number of contacts with the apertures in the screen as the screen loading increases (e.g., material falling onto other material that is already covering the screen apertures), Eq. 2 is modified as follows:

$$N_e = aN_c^{\,b} \qquad (3)$$

where:

N_e = the effective number of contacts with the bare screen surface;

a = a modeling parameter that may be a function of feed rate or holdup; and

b = a modeling parameter that accounts for multiple contacts per aperture.

Both a and b are m x n matrices for an m component and n size class system. It follows that N_e is also an m x n matrix. Thus, the model accounts for a variation in number of contacts as a function of component and of size class.

The probability of passage is modeled as

$$P_o = \frac{(D_{ad} - D_p)(D_{as} - D_p)}{D_{as}\,(D_{ad} + d_d)} \qquad (4)$$

where:

P_o = probability of passage;

D_p = particle size;

D_{ad} = spacing between intermeshed discs;

D_{as} = spacing between shafts; and

d_d = disc thickness.

The current model also incorporates an additional modeling parameter coefficient in the probability of passage relation such that

$$P = cP_o \tag{5}$$

where:

P = the modified probability of passage; and

c = a modeling parameter.

In the model, both P and c are m x n matrices. Therefore, each element in the m x n feed matrix has its own probability of passage associated with it.

The mass fraction of material in a given component size category that reports to the oversize fraction is given by

$$mf_v(i,j) = (1-P)^{N_e}\, mf_f(i,j) \tag{6}$$

where:

$mf_v(i,j)$ = the mass fraction of a given component size category in the overs stream;

$mf_f(i,j)$ = the mass fraction of a given component size category in the feed;

i = the component index; and

j = the size index.

The mass fraction of material in a given component size category that reports to the undersize fraction is given by

$$mf_u(i,j) = (1-(1-P)^{N_e})\, mf_f(i,j) \tag{7}$$

where:

$mf_u(i,j)$ = the mass fraction of a given component size category in the unders stream.

2.2.2 Important Assumptions

The major assumptions in the development of the disc screening model are summarized in Table 2.1.

Table 2.1. Disc Screening Model Assumptions

1. Particles in the layer next to the screen surface encounter each intershaft opening.
2. The material is well-mixed.
3. Oversize material is not forced through the screen by the rotating discs.
4. No size reduction occurs in the screen. Therefore, the sum of the mass of material in a given component size category in the overs and unders streams is equal to the mass of material in that same component size category in the feed.
5. Material entering the screen falls onto the first shaft.
6. The area available for particle passage between the edge of the disc and the neighboring shaft is small relative to the area between intermeshed discs.

2.2.3 Example

Sample calculations for the disc screening model are presented in Figure 2.3. The screen parameters are shown in Table 2.2. No field test data are available for checking (or refining) the model. The overs/unders splits for the sample calculations are 49.8/50.2. The calculated screening efficiency is 70.1 percent.

Table 2.2. Screen Parameters Used in Sample Calculations

Parameter	Value
Feed Rate	3 Mg/hr
Screen Length	3 m
Number of Shafts	10
Shaft Diameter	200 mm
Disc Diameter	300 mm
Shaft Spacing	15.9 mm
Disc Spacing	9.5 mm
Disc Thickness	6.4 mm

Feed

Component	-1.3	-5.1 +1.3	-12.7 +5.1	-25.4 +12.7	+25.4
	Size Class (mm)				
Paper/Plastic	5.41	13.59	21.99	10.30	15.90
Other Organics	5.42	8.25	1.70	0.78	0.85
Glass	5.41	3.89	0.04	0.00	0.00
Other Inorganics	5.42	0.34	0.10	0.08	0.52

↓

Overs Stream

Component	-1.3	-5.1 +1.3	-12.7 +5.1	-25.4 +12.7	+25.4
	Size Class (mm)				
Paper/Plastic	0.01	0.44	18.91	10.30	15.90
Other Organics	0.01	0.27	1.46	0.78	0.85
Glass	0.01	0.13	0.03	0.00	0.00
Other Inorganics	0.01	0.01	0.09	0.08	0.52

+

Unders Stream

Component	-1.3	-5.1 +1.3	-12.7 +5.1	-25.4 +12.7	+25.4
	Size Class (mm)				
Paper/Plastic	5.40	13.15	3.08	0.00	0.00
Other Organics	5.41	7.98	0.24	0.00	0.00
Glass	5.40	3.76	0.01	0.00	0.00
Other Inorganics	5.41	0.33	0.01	0.00	0.00

Figure 2.3. Sample Disc Screening Model Calculations

2.3 ENERGY REQUIREMENTS

The energy requirement for disc screening has been developed using manufacturer's data and operating data for disc screens used in resource recovery processing.

The installed power for two disc screens operating at one resource recovery plant and for disc screens supplied by one manufacturer averages about 0.4 kWh/Mg rated throughput capacity. Invoking the assumption that the motors are used at an average of one-half of their rated output yields an estimated specific energy consumption of 0.2 kWh/Mg.

3. Glass Separation Model

3.1 BACKGROUND

The recovery of glass from municipal solid waste involves the utilization of a number of pieces of processing equipment. Glass separation systems can be divided into two basic technologies, each having associated with it a particular set of unit operations and a particular sequence of processing steps. The first technology, froth flotation, produces a high-purity mixed-color glass of fine particle size (0.1 to 0.9 mm). The technology employs density and size separation, size reduction, and dewatering equipment as well as flotation cells. The second technology yields a color-sorted product of lower purity in the 6 to 19 mm size range. The second technology uses size and density separation equipment prior to optical sorting of the glass concentrate. The froth flotation technology has been selected for modeling inasmuch as it is a prevalent technology and has been reported in the literature.

The glass recovery operation under consideration is depicted in Figure 3.1. The system accepts a -12.5 mm air classified heavy fraction which has undergone one stage of ferrous removal. Material is first processed in a jig to separate the heavy glass-rich fraction from lighter organics. The optimal size range of solid particles for processing in froth flotation cells is 0.1 to 0.9 mm. Accepts from the jig are thus size reduced by crushing in a rod mill after a dewatering stage. Crushed material which passes the openings of a 0.9 mm screen is deslimed (i.e., removal of -0.1 mm material) in a hydrocyclone prior to its introduction to the froth flotation circuit. The hydrophobic nature of a glass particle surface causes such particles to concentrate in the float fraction while the other inerts present in the flotation cells (i.e., tailings which include ceramics, stones, and non-ferrous metals) tend to sink. The glass concentrate is then finally dewatered.

The feed material to the glass recovery operation typically has a glass content of 40 to 60 percent, approximately 25 percent miscellaneous organics with the balance being other inerts. The nominal size of the

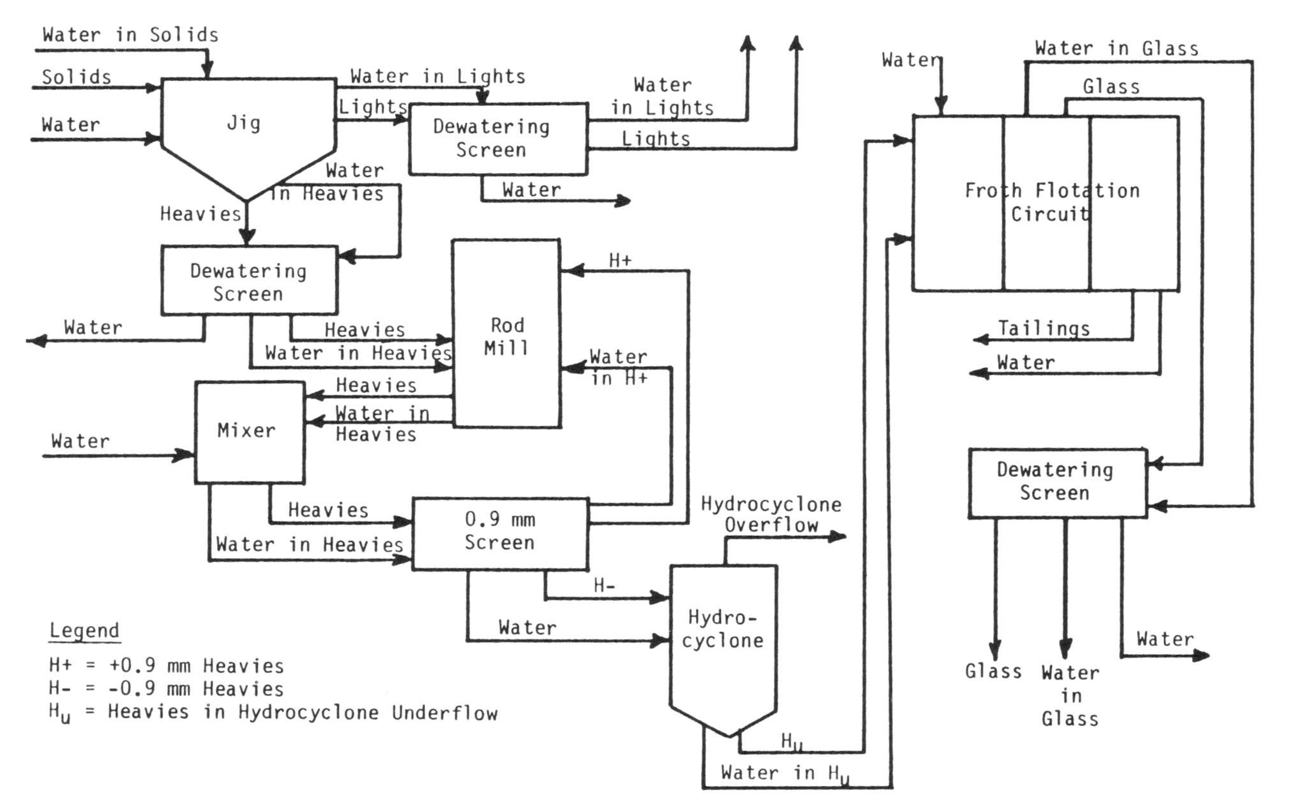

Figure 3.1. Glass Recovery System Flow Diagram

feed ranges from 9 to 11 mm. The glass product has a purity greater than 99 percent and a nominal size between 0.7 and 0.8 mm.

The model of the froth flotation technology has been developed to a large extent through the use of published operating data. Findings from research conducted by the U.S. Bureau of Mines on glass separation from municipal solid waste and incinerator residue by froth flotation were particularly useful in developing the generic model.

The material properties which have been chosen as the principal parameters affecting performance in the model are material mass flowrate, moisture content, and the size and composition of the solids.

3.2 DEVELOPMENT OF MASS BALANCE MODEL

The generic model for glass recovery includes the following six specific unit processes: (1) jigging for density separation; (2) dewatering by screening; (3) crushing; (4) mixing (i.e., slurrying); (5) size separation using a hydrocyclone; and (6) froth flotation for separating glass from other inerts.

Listed in Tables 3.1 through 3.4 are the inputs, transfer functions, and outputs for each unit process.

The jig (cf, Table 3.1) is a split stream process wherein the solids in the feed are considered to have three components, glass, other inerts, and organics (i.e., X_1, X_2, and X_3). The efficiency of the jig in concentrating any particular component in the heavy or light stream ($\underline{\varepsilon}_x$ or $1-\underline{\varepsilon}_x$) has not been reported in the literature. Therefore, for the purposes of the glass recovery model, the efficiency has been estimated. The relationship between the input mass flowrate of a given component, x, ($\underline{\dot{m}}_{x,i}$) and the input mass flowrate of the aggregate solids ($\dot{m}_i$) is shown in Table 3.1. Knowing $\underline{\dot{m}}_{x,i}$ and $\underline{\varepsilon}_x$, the mass flowrates of the components in the output heavy and light streams can be determined (($\underline{\dot{m}}_{x,o})_h$, and $(\dot{m}_{x,o})_l$, respectively). Flowrates of individual components are shown as matrices in which individual size categories are represented.

The size distribution and composition data presented in the discussion of the model are taken from in-house CRS data collected on air classified heavy fraction.

Table 3.1. Jig

Input(s)	Transfer Equation	Output
a) $\dot{m}_i$; $\underline{mf}_{x,i}$	a) $\underline{\dot{m}}_{x,i} = \dot{m}_i \times \underline{mf}_{x,i}$	$\underline{\dot{m}}_{x,i}$
b) $\underline{\dot{m}}_{x,i}$; $\underline{\varepsilon}_x$	b) $(\underline{\dot{m}}_{x,o})_h = \underline{\dot{m}}_{x,i} \times \underline{\varepsilon}_x$	$(\underline{\dot{m}}_{x,o})_h$
c) $\underline{\dot{m}}_{x,i}$; $\underline{\varepsilon}_x$	c) $(\underline{\dot{m}}_{x,o})_l = \underline{\dot{m}}_{x,i} \times (1-\underline{\varepsilon}_x)$	$(\underline{\dot{m}}_{x,o})_l$
d) $\dot{m}_{w,i}$	d) $(\dot{m}_{w,o})_h = \dfrac{\sum_{x=1}^{n} (\dot{m}_{x,o})_h}{\dot{m}_i} \dot{m}_{w,i}$	$(\dot{m}_{w,o})_h$
e) $\dot{m}_{w,i}$	e) $(\dot{m}_{w,o})_l = \dfrac{\sum_{x=1}^{n} (\dot{m}_{x,o})_l}{\dot{m}_i} \dot{m}_{w,i}$	$(\dot{m}_{w,o})_l$

Legend:

$\dot{m}_i$ = Input solids mass flowrate.

$\underline{\dot{m}}_{x,i}$ = Input mass flowrate of component x by size class.

$\underline{\varepsilon}_x$ = Jig effectiveness in concentrating component x in the heavy stream by size class.

$(\underline{\dot{m}}_{x,o})_h$ = Mass flowrate of component x in the heavy output stream by size class.

$(\underline{\dot{m}}_{x,o})_l$ = Mass flowrate of component x in the light output stream by size class.

$\dot{m}_{w,i}$ = Input water mass flowrate.

$(\dot{m}_{w,o})_h$ = Mass flowrate of water in the heavy output stream.

$(\dot{m}_{w,o})_l$ = Mass flowrate of water in the light output stream.

Table 3.2. Dewatering Screen

Input(s)	Transfer Equation	Output
a) $\underline{\dot{m}}_{x,+}$ $\underline{\dot{m}}_{x,-}$ ξ	$\underline{\dot{m}}_{x,ov} = \underline{\dot{m}}_{x,+} + \underline{\dot{m}}_{x,-}(1-\xi)$	$\underline{\dot{m}}_{x,ov}$
b) $\underline{\dot{m}}_{x,-}$ ξ	$\underline{\dot{m}}_{x,un} = \underline{\dot{m}}_{x,-}(\xi)$	$\underline{\dot{m}}_{x,un}$
c) $\underline{\dot{m}}_{x,ov}$; k	$\dot{m}_{w,ov} = \left(\frac{k}{1-k}\right)\left(\sum_{x=1}^{n} \dot{m}_{x,ov}\right)$	$\dot{m}_{w,ov}$
d) $\dot{m}_{w,ov}$; $\dot{m}_{w,i}$	$\dot{m}_{w,un} = \dot{m}_{w,i} - \dot{m}_{w,ov}$	$\dot{m}_{w,un}$

Legend:

$\underline{\dot{m}}_{x,+}$ = Input mass flowrate by size class of component with a particle size larger than the screen aperture.

$\underline{\dot{m}}_{x,-}$ = Input mass flowrate by size class of component x with a particle size smaller than the screen aperture.

ξ = Screening effectiveness.

$\underline{\dot{m}}_{x,ov}$ = Mass flowrate by size class of component x which reports to the the oversize.

$\underline{\dot{m}}_{x,un}$ = Mass flowrate by size class of component x which passes the screen.

$\dot{m}_{w,i}$ = Mass flowrate of water which enters the screen.

$\dot{m}_{w,ov}$ = Mass flowrate of water which reports with the oversize material.

$\dot{m}_{w,un}$ = Mass flowrate of water which passes the screen.

k = Fraction of mass not passing screen which is water.

Table 3.3. Hydrocyclone

Input(s)	Transfer Equation	Output
a) $\underline{\dot{m}}_{x,+}$; r	$\underline{\dot{m}}_{x,un} = \underline{\dot{m}}_{x,+}\ (r)$	$\underline{\dot{m}}_{x,un}$
b) $\underline{\dot{m}}_{x,-}$; $\underline{\dot{m}}_{x,+}$; r	$\underline{\dot{m}}_{x,ov} = \underline{\dot{m}}_{x,-} + (1-r)\ \underline{\dot{m}}_{x,+}$	$\underline{\dot{m}}_{x,ov}$
c) $\dot{m}_{w,i}$; p	$\dot{m}_{w,un} = (p)\ (\dot{m}_{w,i})$	$\dot{m}_{w,un}$
d) $\dot{m}_{w,i}$; p	$\dot{m}_{w,ov} = (1-p)\ \dot{m}_{w,i}$	$\dot{m}_{w,ov}$

Legend:

$\underline{\dot{m}}_{x,+}$ = Input mass flowrate of component x by size class larger than 0.1 mm.

$\underline{\dot{m}}_{x,un}$ = Input mass flowrate of component x by size class in the underflow.

r = Fraction of component x which is larger than 0.1 mm in the underflow.

$\underline{\dot{m}}_{x,-}$ = Mass flowrate of component x by size class which is smaller than 0.1 mm.

$\underline{\dot{m}}_{x,ov}$ = Mass flowrate of component x by size class in the overflow.

$\dot{m}_{w,i}$ = Mass flowrate of water into the hydrocyclone.

$\dot{m}_{w,un}$ = Mass flowrate of water in the underflow.

p = Fraction of water in the input going to the underflow.

$\dot{m}_{w,ov}$ = Mass flowrate of water in the overflow.

Table 3.4. Froth Flotation Circuit

Input(s)	Transfer Equation	Output
a) $\underline{\dot{m}}_{x,+}$; $\underline{\dot{m}}_{x,-}$; q; s	$\underline{\dot{m}}_{x,a} = (q)\underline{\dot{m}}_{x,-} + (s)\underline{\dot{m}}_{x,+}$	$\underline{\dot{m}}_{x,a}$
b) $\underline{\dot{m}}_{x,+}$; $\underline{\dot{m}}_{x,-}$; q; s	$\underline{\dot{m}}_{x,r} = (1-q)\underline{\dot{m}}_{x,-} + (1-s)\underline{\dot{m}}_{x,+}$	$\underline{\dot{m}}_{x,r}$
c) $\dot{m}_{w,i}$; t	$\dot{m}_{w,a} = (t)\ (\dot{m}_{w,i})$	$\dot{m}_{w,a}$
d) $\dot{m}_{w,i}$; t	$\dot{m}_{w,r} = (1-t)\ (\dot{m}_{w,i})$	$\dot{m}_{w,r}$

Legend:

$\underline{\dot{m}}_{x,+}$ = Input mass flowrate of component x by size class entering the circuit which is larger than 0.15 mm.

$\underline{\dot{m}}_{x,-}$ = Mass flowrate of component x by size class entering the circuit which is smaller than 0.15 mm.

q = Fraction of -0.15 mm material which is accepted by the froth flotation cirucit.

s = Fraction of +0.15 mm material which is accepted by the froth flotation circuit.

$\underline{\dot{m}}_{x,a}$ = Mass flowrate of component x which is accepted by the froth flotation circuit.

$\underline{\dot{m}}_{x,r}$ = Mass flowrate of component x which is rejected by the flotation circuit.

$\dot{m}_{w,i}$ = Mass flowrate of water into the circuit.

t = Fraction of input water which goes to the accept stream.

$\dot{m}_{w,a}$ = Mass flowrate of water in the accepts.

$\dot{m}_{w,r}$ = Mass flowrate of water in the rejects.

The incoming mass flowrate of water ($\dot{m}_{w,i}$) is assumed to be split proportionally between the heavy and light streams according to the quantity of solids reported to each.

Table 3.2 contains the relevant inputs, transfer functions, and outputs for a dewatering screen. The mass flowrate of a given component which reports to the oversize ($\underline{\dot{m}}_{x,ov}$) is dependent upon the mass flowrate of that component which is both larger and smaller than the screen aperture ($\underline{\dot{m}}_{x,+}$ and $\underline{\dot{m}}_{x,-}$, respectively), and the screening effectiveness. The amount of water which reports to the oversize stream is related to the quantity of oversize solids by the moisture fraction of the oversize mass flow. For the present model, the moisture fraction is assumed to be constant.

Inert material, which has been separated from organics by the jig and dewatered, is size reduced in a rod mill. The size distribution of the inert material is calculated using the size reduction model developed in Task 3 [1]. The parameters of the model would be chosen to reflect the use of rods as the size reduction elements and the absence of a discharge opening to control the size of the discharged particles.

The transfer functions for the mixer are similar to those for the jig. The basic differences are that there is only one output stream and that an equipment effectiveness value (i.e., the jig's $\underline{\varepsilon}_x$) is not required for the mixer.

Following the mixer, the +0.9 mm and -0.1 mm solids are removed from the input stream to the froth flotation circuit through the use of a vibrating screen and a hydrocyclone, respectively. Since the screen is the same type as that used for the dewatering operation, the equations shown in Table 3.2 apply. The inputs, outputs, and transfer functions for the hydrocyclone are summarized in Table 3.3. Discussion with manufacturers regarding the operation of hydrocyclones disclosed that all of the -0.1 mm ($\underline{\dot{m}}_{x,-}$) material is removed. However, a small fraction of the +0.1 mm particles ($\underline{\dot{m}}_{x,+}$) also reports to the overflow stream. The relations for the underflow ($\underline{\dot{m}}_{x,un}$) and overflow ($\underline{\dot{m}}_{x,ov}$) streams are presented in Table 3.3. The flow of water to the underflow ($\underline{\dot{m}}_{w,un}$) is considered to be a constant fraction (p) of that in the input ($\underline{\dot{m}}_{w,in}$).

Glass is separated from the other inert materials in the underflow stream using a froth flotation circuit. The relations in Table 3.4 describe how the mass flowrate of the accepts, $\underline{\dot{m}}_{x,a}$ (i.e., that which is floated) relates to the solids present in the feed, which have a particle size less than 0.15 mm ($\underline{\dot{m}}_{x,-}$), and those which are larger than 0.15 mm ($\dot{m}_{x,+}$). Data contained in Ref. 2 reveal that the recovery of glass in froth flotation cells in the case of particle sizes less than 0.15 mm (q) is typically 50 percent, while that for sizes greater than 0.15 mm (s) is approximately 95 percent. The quantities of water discharged in the accepted and rejected streams are assumed to be constant fractions of the water fed to the flotation circuit.

3.2.1 Example

The following example serves to illustrate the use of the jig model. The example follows the flow of glass through the jigging operation. The calculations for the other components would be conducted in a similar manner as would the calculations for the subsequent unit operations.

The variables and matrix notation used in the example are described in Table 3.5.

The given information for the jigging example is as follows:

1. Total input solids flow: $\dot{m}_i = 7.0$ Mg/hr
2. Component (x) = glass (gl)
3. $mf_{gl,i} = 0.5$
4. j = 5 size classes
5. $\underline{X}_{gl,i}$ =

Weight Percent	Size (mm)
0.30	+9.53
0.27	-9.53 +6.35
0.23	-6.35 +2.82
0.06	-2.82 +1.42
0.14	-1.42

6. $\underline{\varepsilon}_{gl,j}$ = efficiency

Table 3.5. Variables and Matrix Notation for the Jigging Example

Matrices

$\underline{X}$ = Matrix of mass fraction values by size class.

$\underline{\varepsilon}$ = Matrix of separation efficiency values.

$\underline{mf}$ = Matrix of size class mass fractions.

$\underline{\dot{m}}$ = Matrix of flow rates by size class.

Subscripts

i = input

o = output

x = component

j = size class

Matrix Elements

$\dot{m}_i$ = Solid input flowrate.

$mf_{x,i}$ = Mass fraction of component x in the input.

$X_{j,i}$ = Mass fraction of input material in size class j.

$mf_{x,o}$ = Mass fraction of component x in the output.

$X_{j,o}$ = Mass fraction of output material in size class j.

$\varepsilon_{x,j}$ = Efficiency of device in concentrating component x in the output stream.

$\dot{m}_{x,i}$ = Mass flowrate of component x in the input.

$\dot{m}_{x,o}$ = Mass flowrate of component x in the output stream.

$$= [0.95 \quad 0.95 \quad 0.95 \quad 0.95 \quad 0.95]$$

size class (mm)	+9.53	-9.53 +6.35	-6.35 +2.82	-2.82 +1.42	-1.42

The calculations are as follows:

1. Input mass flowrate of glass ($\underline{\dot{m}}_{gl,i}$)

$$\underline{\dot{m}}_{gl,i} = mf_{gl,i} \times \dot{m}_i \times \underline{X}_{gl,i}$$

$$\underline{\dot{m}}_{gl,i} = 0.5 \times 7.0 \times \begin{pmatrix} 0.3 \\ 0.27 \\ 0.23 \\ 0.06 \\ 0.14 \end{pmatrix} \text{ Mg/hr}$$

$$= \begin{pmatrix} 1.05 \\ 0.95 \\ 0.80 \\ 0.21 \\ 0.49 \end{pmatrix} \text{ Mg/hr}$$

2. Aggregate mass flowrate of glass in the input stream.

$$\dot{\underline{m}}_{gl,i} = \sum_{j=1}^{5} (\dot{m}_{gl,i})_j$$

$$= 1.05 + 0.95 + 0.80 + 0.21 + 0.49 \text{ Mg/hr}$$

$$= 3.50 \text{ Mg/hr}$$

3. Mass flowrate of glass reporting to the heavies stream.

($\underline{\dot{m}}_{gl,o}$)

$$\underline{\dot{m}}_{gl,o} = \dot{m}_{gl,i} \times \underline{\varepsilon}_{gl,i}$$

$$\underline{\dot{m}}_{gl,o} = \begin{pmatrix} 1.05 \\ 0.95 \\ 0.80 \\ 0.21 \\ 0.49 \end{pmatrix} \times [0.95 \quad 0.95 \quad 0.95 \quad 0.95 \quad 0.95] \text{ Mg/hr}$$

$$= \begin{pmatrix} 1.00 \\ 0.90 \\ 0.76 \\ 0.20 \\ 0.47 \end{pmatrix} \text{ Mg/hr}$$

4. Aggregate mass flowrate of glass in the heavies stream.

$$\dot{m}_{gl,o} = \sum_{j=1}^{5} (\dot{m}_{gl,o})_j$$

$$= 1.00 + 0.90 + 0.76 + 0.20 + 0.47 \text{ Mg/hr}$$

$$= 3.33 \text{ Mg/hr}$$

5. Size distribution of glass in the heavies stream ($\underline{X}_{gl,o}$)

$$\underline{X}_{gl,o} = \underline{\dot{m}}_{gl,o} / \sum_{j=1}^{5} (\dot{m}_{gl,o})_j$$

$$= \begin{pmatrix} 1.00 \\ 0.90 \\ 0.76 \\ 0.20 \\ 0.47 \end{pmatrix} \div 3.33$$

$$= \begin{pmatrix} 0.30 \\ 0.27 \\ 0.23 \\ 0.06 \\ 0.14 \end{pmatrix}$$

3.3 ENERGY REQUIREMENTS

The energy relations for each unit process included in the glass recovery system are listed in Table 3.6. A standard power equation for pumps has been used for the jig and the hydrocyclone for the purpose of establishing the specific energy for the two operations. In the case of the dewatering screen, mixer, and froth flotation circuit, energy usage data supplied by manufacturers was used to formulate the power relations.

The required inputs for jig and hydrocyclone energy models are solid and water mass flowrates, total dynamic pump head (H_p), and pump

Table 3.6. Glass Recovery System Energy Relations

Unit Process	Input(s)	Power Relation	Specific Energy
Jig	$\dot{m}_w = 21$ Mg/hr $\sum_x \dot{m}_x = 7$ Mg/hr $\dot{m}_w$ = mass flowrate of water $\sum_x \dot{m}_x$ = total solids mass flowrate	$P = \frac{\dot{W} H_p}{\eta_p}$ $= 0.5$ kW $\dot{W}$ = water flowrate by weight (assumed to be 0.06 kN/sec) H_p = total head requirement (assumed to be 6 m) η_p = pump efficiency (assumed to be 0.7)	$E_o = \frac{P}{\dot{M}_T{}^a}$ $= 0.02$ kWh/Mg $\dot{M}_T^a = 28$ Mg/hr
Dewatering Screen	$\dot{M}_T = 14$ Mg/hr	$P = 0.2$ kW[b]	$E_o = \frac{P}{\dot{M}_T}$ $= 0.01$ kWh/Mg
Mixer	$\dot{M}_T = 15$ Mg/hr	$P = 1.7$ kW[b]	$E_o = \frac{P}{\dot{M}_T}$ $= 0.11$ kWh/Mg

Table 3.6 (Cont'd)

Unit Process	Input(s)	Power Relation	Specific Energy
Hydrocyclone	$\dot{m}_w = 9$ Mg/hr $\sum_x \dot{m}_x = 3$ Mg/hr	$P = \frac{\dot{W}_T H_p}{\eta_p}$ $\simeq 0.3$ kW $\dot{W}_T$ = total (water + solids) weight flowrate (assumed to be 0.03 kN/sec) $H_p = 6$ m $\eta_p = 0.7$	$E_o = \frac{P}{\dot{M}_T}$ $= 0.025$ kWh/Mg $\dot{M}_T = 12$ Mg/hr
Froth Flotation Circuit	$\dot{M}_T = 10$ Mg/hr	$P = 4$ kW[b]	$E_o = \frac{P}{\dot{M}_T}$ $= 0.4$ kWh/Mg

[a] $\dot{M}_T$ = Total mass flowrate, i.e., solids plus water.

[b] Value obtained from manufacturer's data.

efficiency (η_p). Solid and water mass flowrates only are required for the mixer, dewatering screen, and froth flotation circuit.

Specific energy consumptions range from 0.02 kWh/Mg (jig) to 0.4 kWh/Mg (froth flotation circuit).

3.4 REFERENCES

1. Cal Recovery Systems, Task 3 Report, Size Reduction Model, ANL Contract No. 31-109-38-7167

2. Heginbotham, J.H., "Recovery of Glass from Urban Refuse by froth Flotation," U.S. Bureau of Mines, College Park, Md., 1978.

4. Non-Ferrous Separation Model

4.1 BACKGROUND

Aluminum is the major non-magnetic metal typically found in municipal solid waste. Aluminum accounts for about 90 percent (by weight) of the total non-magnetic metals in the waste stream. However, relative to the entire waste stream, non-magnetic metals constitute only about 1 to 2 percent (by weight).

The existing systems designed to recover aluminum from mixed municipal solid waste require that the material undergo a series of preliminary steps before the waste is introduced to the aluminum recovery unit. The various pre-processing steps are designed to achieve a certain degree of concentration of the non-ferrous (aluminum) fraction of the waste stream. Generally, one of the first preparatory steps is size reduction. Usually size reduction is followed by air classification, magnetic separation, and, in some cases, screening.

The majority of the full-scale aluminum separation systems installed as part of a material recovery process have not operated successfully in the past. Furthermore, detailed evaluations of their operation have not been carried out. The few analyses available in the literature usually are general in nature and contain little, if any, technical information.

Of the technologies that have been used for non-ferrous separation, the eddy current technology appears to be the most utilized and the most feasible. Consequently, the model presented here is based on the eddy current technology.

The eddy current separation units perform material segregation through the use of a time-varying magnetic field, which induces small electrical currents and, therefore, repulsive forces on the non-ferrous materials. The magnitude of the repulsive force is a function of the magnitude of the magnetic field and of the properties of the non-ferrous particles (e.g., geometry, size, conductivity).

Most of the devices designed to recover aluminum from the waste stream generate the source magnetic field in either of two ways. In the first method, the aluminum particle moves through a magnetic field generated by an array of permanent magnets. In the second approach, the magnetic field is varied by means of an a-c electromagnetic system. Linear motors also have been used to recover aluminum from municipal solid waste.

Unfortunately, the aluminum separation devices are not sufficiently sensitive to differentiate between aluminum and other non-ferrous metals. Furthermore, as the non-ferrous metals are deflected from the separation unit, they generally become entwined with contaminants such as organic matter and textiles that may be in their path. These contaminants must eventually be removed since their concentration generally is higher than that allowed by the aluminum scrap buyers.

4.2 DEVELOPMENT OF MASS BALANCE MODEL

The model for the aluminum recovery unit has been developed based on information available in the literature. The model consists of three specific subsystems designed to gradually concentrate the aluminum fraction. The subsystems are shown in Figure 4.1. The figure shows that the overall system consists of three components: a trommel screen, an aluminum magnet, and an air knife. The trommel screen is designed such that the feed is separated into three size classes. The fraction containing the highest concentration of aluminum becomes the feed into the aluminum magnet. The "accepts" or aluminum concentrate from the aluminum magnet is further treated in an air knife in order to remove organic and other non-ferrous contaminants.

In the model, the inputs (feedstock) are described by matrices. Each input into a particular subsystem is operated on by a transfer function, also in the form of a matrix, which describes the separation efficiency of the particular device. The output from each unit also is described in the form of a matrix. Therefore, the ability of a particular device to segregate a certain material from the waste stream can be expressed as follows:

$$\underline{mf}_i \times \underline{\varepsilon} = \underline{mf}_o$$

A list of variables used in describing each system is given in Table 4.1.

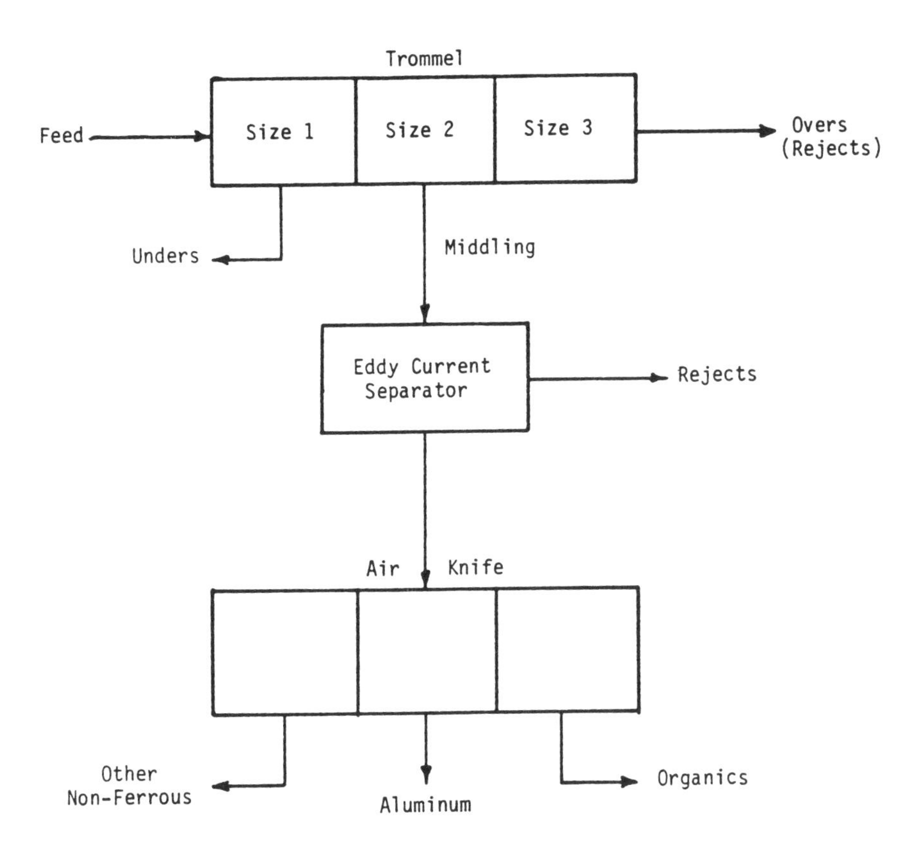

Figure 4.1. Schematic Diagram of Aluminum Recovery System Using an Eddy Current Separator

Table 4.1. List of Variables

$\dot{M}$	= mass flowrate (total)
Matrix Elements	
$\underline{mf}$	= matrix of mass fractions
$\underline{\dot{m}}$	= matrix of mass flowrates
$\underline{\alpha}, \underline{\varepsilon}, \underline{\beta}$	= matrix of separation efficiencies
Subscripts	
x	= component
pp	= paper, plastic
Al	= aluminum
Fe	= ferrous metals
Ot	= other
i	= input
o	= output

Since the first separation device in the overall system is a trommel screen, which segregates the feed stream into three size classes, a description of the size distribution by component would be superfluous.

It is important to note that the fraction of the waste stream being processed is described in terms of four components or fractions: (1) paper and plastic; (2) aluminum; (3) ferrous metals; and (4) other. Each one of these materials is accounted for as it passes through the various subsystems.

The variables used in the models of the subsystems are described in Tables 4.2 to 4.4. Representative values for the component separation efficiencies are also indicated in the tables for each subsystem.

Table 4.2. Description of Variables for Aluminum Magnet

Input	Transfer Function	Output (Accepts)
$mf_{x,i}$	$\varepsilon_{x,j}$	$mf_{x,o}$
$\begin{pmatrix} mf_{pp,i} \\ mf_{Al,i} \\ mf_{Fe,i} \\ mf_{Ot,i} \end{pmatrix}$	$\begin{pmatrix} 0.089 \\ 0.636 \\ 1.0 \\ 0.083 \end{pmatrix}$	$\begin{pmatrix} mf_{pp,o} \\ mf_{Al,o} \\ mf_{Fe,o} \\ mf_{Ot,o} \end{pmatrix}$

Table 4.3. Description of Variables for Trommel Screen

Input	Transfer Functions			Output		
	Size 1 (-5/8")	Size 2 (+5/8"-2")	Size 3 (+2")	Size 1	Size 2	Size 3
$mf_{x,i}$	$\alpha_{x,j}$	$\varepsilon_{x,j}$	$\beta_{x,j}$	$mf_{x,o}$	$mf_{x,o}$	$mf_{x,o}$
$mf_{pp,i}$	0.547	0.247	0.206	$mf_{pp,o1}$	$mf_{pp,o2}$	$mf_{pp,o3}$
$mf_{Al,i}$	0.10	0.553	0.347	$mf_{Al,o1}$	$mf_{Al,o2}$	$mf_{Al,o3}$
$mf_{Fe,i}$	0.50	0.083	0.417	$mf_{Fe,o1}$	$mf_{Fe,o2}$	$mf_{Fe,o3}$
$mf_{Ot,i}$	0.677	0.124	0.199	$mf_{Ot,o1}$	$mf_{Ot,o2}$	$mf_{Ot,o3}$

Table 4.4. Description of Variables for Air Knife

Input	Transfer Functions			Output		
	Other Non-Fe	Al	Organics	Other Non-Fe	Al	Organics
$mf_{x,i}$	$\alpha_{x,j}$	$\epsilon_{x,j}$	$\beta_{x,j}$	$mf_{x,o}$	$mf_{x,o}$	$mf_{x,o}$
$mf_{pp,i}$	0.18	0.019	0.80	$mf_{pp,o1}$	$mf_{pp,o2}$	$mf_{pp,o3}$
$mf_{Al,i}$	0.70	0.29	0.0	$mf_{Al,o1}$	$mf_{Al,o2}$	$mf_{Al,o3}$
$mf_{Fe,i}$	0.70	0.3	0.0	$mf_{Fe,o1}$	$mf_{Fe,o2}$	$mf_{Fe,o3}$
$mf_{Ot,i}$	0.37	0.02	0.61	$mf_{Ot,o1}$	$mf_{Ot,o2}$	$mf_{Ot,o3}$

The matrix used to describe the mass fraction input information ($\underline{mf}$) to a particular device takes the following form:

$$\underline{mf} = \begin{pmatrix} mf_{pp,i} \\ mf_{Al,i} \\ mf_{Fe,i} \\ mf_{Ot,i} \end{pmatrix} \quad (1)$$

The input matrix is multiplied by the matrix containing the values for separation efficiencies for each material. A separation efficiency matrix ($\underline{\eta}$) is written as follows:

$$\underline{\eta} = \begin{pmatrix} \varepsilon_{pp,j} \\ \varepsilon_{Al,j} \\ \varepsilon_{Fe,j} \\ \varepsilon_{Ot,j} \end{pmatrix} \quad (2)$$

The multiplication of matrices (1) and (2) yields a matrix which describes the output of each device. This matrix takes the following form:

$$\underline{mf} = \begin{pmatrix} mf_{pp,o} \\ mf_{Al,o} \\ mf_{Fe,o} \\ mf_{Ot,o} \end{pmatrix} \quad (3)$$

The following example demonstrates the concentration of aluminum by an eddy current separator. The basic mathematical expression is:

$$\underline{mf}_i \times \underline{\varepsilon} = \underline{mf}_o$$

Typical mass fractions for paper and plastic, aluminum, ferrous, and other materials are given as:

$$\begin{pmatrix} 0.637 \\ 0.057 \\ 0.005 \\ 0.301 \end{pmatrix} \times (0.089 \quad 0.636 \quad 1.0 \quad 0.083) = \begin{pmatrix} 0.057 \\ 0.036 \\ 0.005 \\ 0.025 \end{pmatrix}$$

Using a total mass flowrate ($\dot{M}$) of 4 Mg/hr, and the following expression:

$$\dot{m}_{Al,o} = mf_{Al,o} \times \dot{M} \tag{4}$$

The amount of aluminum reporting to the accepts of the separator can be calculated,

$$\dot{m}_{Al,o} = 0.036 \times 4.0 = 0.144 \text{ Mg/hr} \tag{5}$$

This example also serves to demonstrate the fact that the concentrate from the separator still contains a certain amount of impurities. For instance the quantity of paper and plastics in the accepts is:

$$\dot{m}_{pp,o} = 0.057 \times 4.0 = 0.228 \text{ Mg/hr} \tag{6}$$

4.3 ENERGY REQUIREMENTS

The power requirement for the non-ferrous separation system is equal to the sum of the power requirements for the three major pieces of equipment in the system, namely, screen, eddy current separator, and air knife. The specific energy is given by

$$E_o = \sum_{i=1}^{3} E_{oi} mf_i \tag{7}$$

where:

$i = 1$ for the screening;

$i = 2$ for the eddy current separator; and

$i = 3$ for the air knife.

The values of E_{oi} and mf_i are typically as follows:

$E_{o1} = 0.4$ kWh/Mg $\qquad mf_1 = 1.0$

$E_{o2} = 2$ kWh/Mg $\qquad mf_2 = 0.2$

$E_{o3} = 5$ kWh/Mg $\qquad mf_3 = 0.03$

The specific energy requirement of the non-ferrous separation is therefore about 1 kWh/Mg of feedstock to the system.

4.4 REFERENCES

1. Ralston, O.C., Electrostatic Separation of Mixed Granular Solids, Elsevier, Amsterdam, 1961.

5. Economic Model

5.1 INTRODUCTION

An economic model for the unit operations in a resource recovery system is presented in this section. The section is organized according to cost component (i.e., capital, operation, maintenance, residue disposal, and revenue from the sale of materials) with the unit operations (shear shredding, disc screening, glass separation, and non-ferrous material separation) discussed as sub-categories under each cost component. The reason for this method of organization is that the portion of each cost component that is attributable to each unit operation often is not determinable when the unit operation is considered in isolation from the rest of the system.

The capital and maintenance cost equations primarily are based on price estimates and quotes from equipment suppliers, from published data on existing resource recovery operations, and from information obtained during interviews with managers of existing operations. In a few cases, estimates of the labor required to maintain equipment are based on CRS's experience. A wage rate of $30/hr is assumed in these instances. Maintenance costs include both labor and supplies.

The base cost of the equipment (i.e., uninstalled) is adjusted by the amounts indicated in Table 5.1 to arrive at an installed cost.

The operating labor requirements are estimated on the basis of engineering judgment. The energy requirements are an input to the economic model. They are determined from the mass and energy balances described previously in this report.

The revenue from the sale of recovered materials and the cost of disposing residue are derived from the mass of recovered material and of residue, both of which are calculated in the mass balance model and are, thus, inputs to the economic model.

Design inputs and site-specific cost inputs that must be supplied by the user of this model are noted in the text as the need for them arises.

Table 5.1. Adjustments to Base Costs of Equipment

Cost Item	Equipment	Adjustment[a] (%)
Engineering services	All	10
Contractor overhead and profit	All	10
Instrumentation, controls, and other accessory equipment	All	20
Sales tax	All	5
Transportation	All	5
Foundations	All	5
Installation	Shear shredder	10
	Disc screen	10
	Glass separation	20
	Non-ferrous separation	10

[a] Adjustments are given as percentages of the uninstalled equipment cost.

The accuracy of the model is estimated to be as follows:

1. Capital costs - ±20 percent
2. Operating costs
    a. Labor - ±30 percent
    b. Energy - +30 percent, -10 percent
3. Maintenance costs - ±30 percent
4. Disposal costs - accuracy is limited by the accuracy of the mass balance model
5. Revenue - accuracy is limited by the accuracy of the mass balance model

The requirement for operating labor depends on the design and degree of automation of the system. The associated error is estimated to be within the range of ±30 percent. The maintenance cost data provided by equipment suppliers varied widely; and in some cases, there is little or no operating experience upon which to base a cost estimate. Consequently, the maintenance costs estimated by the model are considered accurate within the range of ±30 percent.

An important limitation on the accuracy of the model derives from the assumption that equipment is operated at its rated capacity. The underutilization of equipment can result in a significantly higher unit cost of capital than what the model predicts. Other costs may also be affected to a smaller degree.

The model is valid for the following mass throughput rates:

1. Shear shredding - >15 Mg/hr
2. Disc screening - >15 Mg/hr
3. Glass separation - 5-10 Mg/hr
4. Non-ferrous separation - 20-80 Mg/hr

In cases where an upper limit to the throughput rate is given, the costs for higher rates may be estimated by assuming that the unit cost for the higher rate is equal to the unit cost for the maximum rate specified above. The limits on the throughput rates mainly arise from the trends evident in the capital cost for various sizes of equipment. Upper limits on the throughput rates occur for unit operations for which there are significant economies of scale within the given range of throughputs.

Costs that are not included in the model include:

1. Building and site improvement
2. Utility connections to the plant site
3. Supervisory, clerical, janitorial, and other indirect labor
4. Transportation of salable materials and residue except to the extent that the user of the model includes these costs in the unit price of recovered material and the unit cost of residue disposal
5. Loading raw MSW into the initial unit operation
6. Storage of MSW, residue, and recovered materials

5.2 CAPITAL COSTS

The models described herein represent the capital costs (in 1984 dollars) of refuse processing unit operations. The models for each unit operation (i.e., shear shredding, etc.) are developed and reported in separate subsections. A summary of the relations for the capital costs of the various unit operations is given in Table 5.2.

Table 5.2. Capital Costs of Unit Operations

Unit Operation	Type of Equipment	Cost ($)	Range of Throughputs (Mg/hr)
Primary Size Reduction	Shear Shredder	12,100 $\dot{m}$	> 15
Secondary Size Reduction	Shear Shredder	12,100 $\dot{m}$	> 15
Screening	Disc Screen	3600 $\dot{m}$	> 15
Glass Separation	Jig, Froth Flotation, etc.	130,000 + 24,000 $\dot{m}$	5 to 10
Non-Ferrous Material Recovery	Screen, Eddy Current Separator, and Air Knife	26,200 $\dot{m}$ - 59.5 $\dot{m}^2$	20 to 80

5.2.1 Shear Shredding

Shear shredder prices were quoted by three manufacturers. While prices given by individual firms indicated significant differences between the cost of primary shredding and secondary shredding, when the prices given by all the manufacturers were considered, the typical price of a primary shredder was not significantly different from that of a secondary shredder. The typical installed price (i.e., the base price plus the adjustments given in Table 5.1) in both cases is

$$C_{ss} = 12{,}100\ \dot{m}_{ss} \qquad (1)$$

where:

C_{ss} = installed cost of shear shredder ($); and
$\dot{m}_{ss}$ = rated mass throughput rate of the shear shredder (Mg/hr).

5.2.2 Disc Screening

Most manufacturers of disc screens are unable to recommend a particular size or model for a given feedrate of MSW or processed MSW. However, costs were obtained from two resource recovery plants that utilize disc screens. They were the Ames, Iowa, plant and the Baltimore County plant. When adjusted for inflation and installation the cost of screens in both plants was about

$$C_{ds} = 3600\ \dot{m}_{ds} \qquad (2)$$

where:

C_{ds} = installed cost of disc screen ($); and
$\dot{m}_{ds}$ = rated mass throughput of the disc screen (Mg/hr).

5.2.3 Glass Separation

The cost of the major pieces of equipment shown in Figure 3.1 plus the cost of material transport equipment (i.e., conveyors, pumps, pipe, etc.) is about $142,000 for a 5 Mg/hr system and $210,000 for a 10 Mg/hr system. The throughput rates are based on the mass of the -12 mm air classified heavy fraction that is the feedstock to the jig. The price of each major piece of equipment was obtained from manufacturers or their representatives. The price of material handling equipment was obtained from industrial supply catalogs. Although conveying is a separate unit process discussed separately in this report, the cost of conveying material within the glass separation system is included in the cost of the system.

An adjustment of 75 percent is used in calculating the installed cost of the glass recovery system. It is higher than the adjustment factor used in the rest of this report because installation is assumed to cost 20 percent rather than 10 percent of the base cost. The higher value is a consequence of the necessity to install several pieces of equipment including pumps, pipes, and conveyors.

The installed cost is given by

$$C_g = 130{,}000 + 24{,}000\, \dot{m}_g \tag{3}$$

where:

C_g = installed cost of glass separation system (\$); and
$\dot{m}_g$ = rate at which undersize air classified heavy fraction enters the system (Mg/hr).

5.2.4 Non-Ferrous Separation

One manufacturer quoted prices of three sizes of eddy current separators that, when adjusted by 65 percent, correspond to the equation

$$C_e = 86{,}000\, \dot{m}_e - 1100\, \dot{m}_e^2 \tag{4}$$

where:

C_e = cost of eddy current separator (\$); and
$\dot{m}$ = rated capacity of the eddy current separator (Mg/hr).

Since the feedstock to the eddy current separator is about 20 percent of the feedstock to the aluminum separation system, Eq. 4 can be rewritten as

$$C_e = 17{,}200\, \dot{m}_a - 44\, m_a^2 \tag{5}$$

where:

$\dot{m}_a$ = feedrate to aluminum separation system.

The cost of the screen in the aluminum separation system is predicted by

$$C_s = 8600\ \dot{m}_a - 13\ \dot{m}_a^2 \tag{6}$$

This equation is developed in the Task 3 Report.

The cost of the air knife is taken to be the same as that of an air classifier which is given in the Task 3 Report as

$$C_{ac} = 16{,}000\ \dot{m}_{ac} - 100\ \dot{m}_{ac}^2 \tag{7}$$

Since only one thirtieth of the feedstock to the aluminum recovery system reaches the air knife, the cost is

$$C_k = 500\ \dot{m}_a - 3\ \dot{m}_a^2 \tag{8}$$

where:

C_k = cost of air knife ($).

The entire cost of the aluminum separation system (C_a) is the sum of the values given in Eqs. 5, 6, and 8.

$$C_a = 26{,}300\ \dot{m}_a - 60\ \dot{m}_a^2 \tag{9}$$

5.3 OPERATING COSTS

The operating costs are composed of labor costs and energy costs. The unit processes discussed in this report use electrical energy only. Usually, no operating supplies are required. Materials and supplies required for maintenance are included in the maintenance cost model.

In the cost model the labor requirements are determined for the entire processing system rather than for each individual unit operation, except in the case of glass separation. The reason is that the labor required for a unit operation depends to a large extent on the configuration of the entire system. Furthermore, the allocation of labor among pieces of equipment is difficult to model.

The model for labor is valid for MSW processing systems with throughput rates exceeding 10 Mg/hr. (Individual unit operations may have throughput rates of less than 10 Mg/hr.) It is designed to account for multiple processing lines. Furthermore, it is designed to be compatible with the labor cost model presented in the Task 4 Report. That is, the labor requirements of a system comprising unit operations from the Task 4 Report and unit operations from this report can be determined using this model.

The model for energy costs requires inputs from the energy requirement models.

5.3.1 Labor

The unit cost of labor is given by the equation,

$$UCL = \frac{C_L}{\dot{m}_1} \tag{10}$$

where:

UCL = unit cost of labor (\$/Mg);
C_L = cost of labor to operate the plant (\$/hr); and
$\dot{m}_1$ = mass flowrate of raw MSW.

The hourly cost of labor (C_L) is equal to the product of the wage rate and the number of machine operators and spotters, i.e.,

$$C_L = W_m \sum_{k=1}^{p} N_{mk} + W_s \sum_{k=1}^{p} N_{sk} \tag{11}$$

where:

W_m = fully burdened wage rate for equipment operators;
W_s = fully burdened wage rate for spotters;
N_m = number of equipment operators in plant;
N_s = number of spotters in plant;
k = group of unit operations; and
p = number of unit operation groups in plant.

Equipment operators are those who control machinery from a control room and monitor operations via closed circuit television. Spotters monitor, and sometimes control, processes in situ and have verbal contact with the equipment operator(s) via radio or telephone.

The unit operation groups (UOG's) comprise one or more unit operations, and each operation includes one or more processing lines. The number of lines in each unit operation is selected by the designer. The UOG's are as follows:

UOG 1 (k=1) comprises unit operations other than conveying that receive raw MSW.

UOG 2 (k=2) comprises any one of the following unit operations that immediately follows UOG 1: size reduction, screening, and non-ferrous separation. If unit operations discussed in the Task 4 Report are included in the system, air classification could also be included in UOG 2.

UOG 3 (k=3) includes any two of the unit operations specified under UOG 2 that are not included in UOG 1 or UOG 2. It should be noted that a unit operation may occur at more than one place in the processing system and is considered a separate unit operation each time it occurs.

UOG 4,5,6,etc. (k=4,5,6,etc.) include three of the unit operations specified under UOG 2 that are not included in a previous UOG.

UOG p (k=p) includes only glass separation. If unit operations discussed in the Task 4 Report are included, densification would also be included in UOG p.

Figure 5.1 illustrates the grouping of unit operations into unit operation groups in a hypothetical processing plant. In some instances, the assignment of a unit operation to a particular UOG is arbitrary. For example, shear shredder 2 could have been assigned to UOG 3 rather than to UOG 4; and secondary screening of non-ferrous separation could have been included in UOG 4 rather than in UOG 3.

In the operating cost model, the wage rates (W_m and W_s in Eq. 11) are inputs to the model as are the number of UOG's (p).

For the three unit operations, size reduction, screening, and non-ferrous separation, the operating labor requirement is calculated by first computing a 'use factor' as follows

$$A_k = \sum_{i=1}^{n_k} \frac{\dot{m}_{ki}}{\dot{m}_{ki}+10} \qquad (12)$$

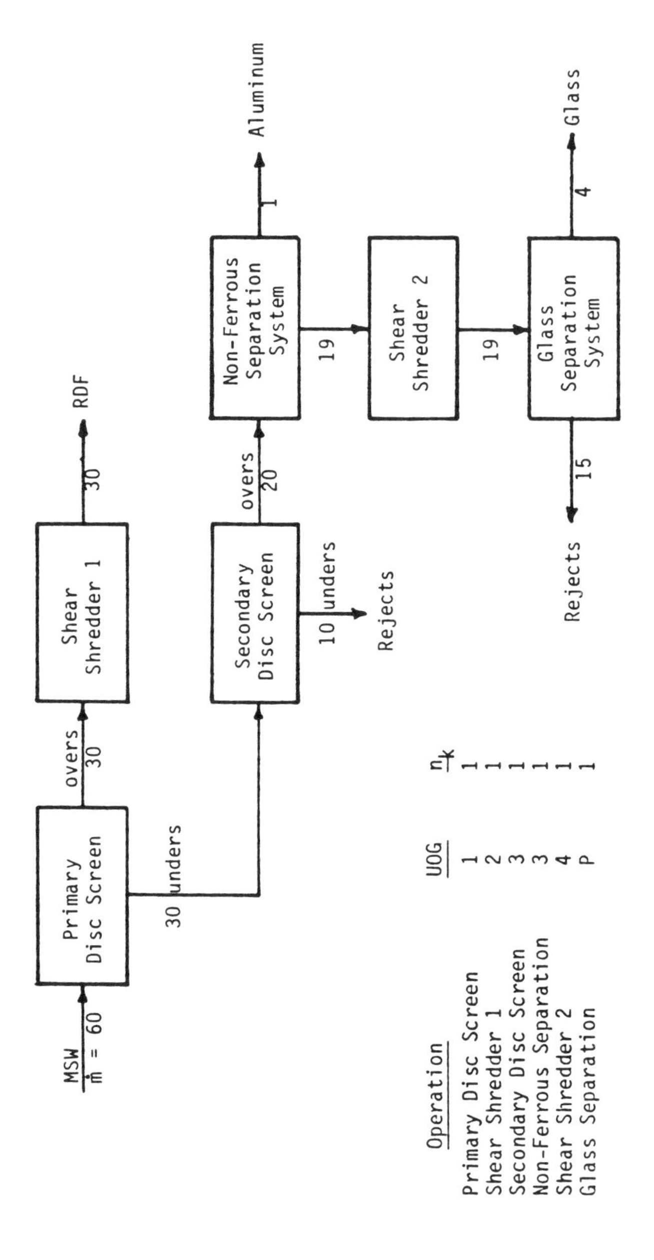

Operation	UOG	n_k
Primary Disc Screen	1	1
Shear Shredder 1	2	1
Secondary Disc Screen	3	1
Non-Ferrous Separation	3	1
Shear Shredder 2	4	1
Glass Separation	P	1

Figure 5.1. Illustration of Nomenclature Used in Determining Labor Requirements

where:

A_k = use factor for UOG k;

$\dot{m}_{ki}$ = mass throughput rate (Mg/hr) for processing line i of UOG k; and

n_k = number of processing lines in UOG k.

For the system shown in Figure 5.1, the use factor for unit operation group 3 is calculated as follows (note that n_k equals two because there are two processing lines, one for secondary screening and one for non-ferrous separation):

$$A_3 = \sum_{i=1}^{2} \frac{\dot{m}_{3i}}{\dot{m}_{3i}+10} \tag{13}$$

$$= \frac{30}{30+10} + \frac{20}{20+10}$$

$$= 1.42$$

By similar calculations, the values of A_1, A_2, and A_4 are determined,

$A_1 = 0.86$

$A_2 = 0.75$

$A_4 = 0.66$

Except when $A_1 + A_2 < 1$, the number of machine operators and spotters is determined by

$$N_{Tk} = \text{INT}\left(\frac{A_k}{2} + 1\right) \tag{14}$$

where:

N_{Tk} = number of machine operators plus number of spotters required for UOG k.

INT indicates that only the integer part of the quantity in brackets is taken. That is, the value in brackets is rounded downward to a whole number.

If $A_1 + A_2 < 1$, then N_{T2} equals zero and N_{T1} is calculated from Eq. 14. Letting

N_{mk} = number of machine operators required for UOG k, and

N_{sk} = number of spotters required for UOG k

and noting that

$$N_{Tk} = N_{mk} + N_{sk}$$

the number of machine operators and spotters is given by

$$N_{Tk} = N_{mk} \text{ when } k=1 \tag{15a}$$

$$N_{mk} = \text{INT}\left(\frac{N_{Tk}}{2}\right) \text{ when } k=2 \text{ or } p>k>3 \tag{15b}$$

$$N_{sk} = \text{INT}\left(\frac{N_{Tk}}{2}\right) \text{ when } k=3 \tag{15c}$$

Using Eqs. 14 and 15 to determine the labor requirements for the system shown in Figure 5.1 (excluding glass separation) yields,

$$N_{T1} = \text{INT}\left(\frac{0.86}{2} + 1\right) = \text{INT}\ (1.43) = 1$$

$$N_{T2} = 1$$

$$N_{T3} = 1$$

$$N_{T4} = 1$$

$$N_{m1} = 1;\ N_{s1} = 0$$

$$N_{m2} = 0;\ N_{s2} = 1$$

$$N_{m3} = 1;\ N_{s3} = 0$$

$$N_{m4} = 0;\ N_{s4} = 1$$

Thus, two equipment operators and two spotters are required.

The labor requirement for glass separation (k=p) is founded on the estimation that one spotter is required for a glass separation system in any MSW processing plant that can be reasonably anticipated. Thus

$$N_{Tp} = N_{sp} = 1 \tag{16}$$

The entire system shown in Figure 5.1, then, requires two equipment operators in the control room and three spotters. Assuming a fully burdened wage rate of $25 per hour for both equipment operators and spotters yields, from Eq. 11,

$$C_L = 25(2) + 25(3)$$
$$= \$125/\text{hr}$$

Substituting this value in Eq. 10 and noting that the mass input to the system is 60 Mg/hr yields a unit cost for labor of $2.08/Mg.

5.3.2 Energy

The specific energy requirements (i.e., energy per unit mass of feedstock to a given unit process) for each unit process (E_o) are modeled in a preceeding section of this report and are summarized in Table 5.3. All of the energy is in the form of electricity.

The unit cost of energy for unit operation j is

$$UCE_j = PE_{oj} \tag{17}$$

where:

UCE_j = unit cost of energy for unit operation j ($/Mg);
P = price of electrical energy ($/kWh); and
E_{oj} = specific energy requirement for unit operation j.

The price is an input to the model.

The unit cost of energy for the entire processing system is given on the basis of a unit mass of MSW feedstock to the plant,

$$UCE_s = \frac{\sum_j (UCE_j \, \dot{m}_j)}{\dot{m}_s} \tag{18}$$

where:

UCE_s = unit cost of energy for the processing system ($/Mg);
$\dot{m}_j$ = mass throughput rate to unit process j (Mg/hr); and
$\dot{m}_s$ = mass throughput rate to the processing system (Mg/hr).

5.4 MAINTENANCE COSTS

The unit cost of maintenance for each unit operation is given in the following section. The costs are summarized in Table 5.4.

5.4.1 Shear Shredding

One manufacturer of shear shredders estimated the unit maintenance cost of a shear shredder to be about $0.80/Mg, 90 percent of which is for

Table 5.3. Specific Energy Requirements

Unit Operation	Equipment	Specific Energy (kWh/Mg)
Size Reduction	Shear Shredder	$\left(\sum_{i=1}^{n} \dot{m}_i A_i Z_{oi}^{b_i} \right) \Big/ \sum_{i=1}^{n} \dot{m}_i$
Screening	Disc Screen	0.2
Glass Separation	Jig, Froth Flotation, etc.	1
Non-Ferrous Separation	Screen, Eddy Current Separator, and Air Knife	1

Table 5.4. Unit Costs of Maintenance

Unit Operation	Type of Equipment	Unit Cost ($/Mg)
Size Reduction (Primary and Secondary)	Shear Shredder	0.80
Screening	Disc Screen	0.20
Glass Separation	Jig, Froth Flotation, etc.	$0.4 + 2/\dot{m}$
Non-Ferrous Material Recovery	Screen, Eddy Current Separator, and Air Knife	0.30

blade maintenance. A second manufacturer estimated total operating and maintenance costs of up to $3/Mg for a 50 Mg/hr shredder. The $3/Mg value includes depreciation, direct and indirect labor, power, parts, repairs, and cutter reconditioning. Cutter reconditioning alone was estimated by the manufacturer to cost $0.66 to $0.88 per Mg which is in rough agreement with the estimate for blade maintenance given above.

The data collected from manufacturers did not permit a differentiation to be made between the unit cost of maintaining a shredder used for primary size reduction and one used for secondary size reduction. Using the available data, the maintenance cost for shear shredding is estimated to be $0.80/Mg.

5.4.2 Disc Screening

No maintenance cost data were available from disc screen manufacturers. There are two disc screens at the Ames resource recovery plant for which maintenance costs are available. Based upon cost data supplied by the plant personnel and CRS engineering judgment, the unit cost of maintenance for disc screens is estimated as $0.20/Mg.

5.4.3 Glass Separation

Maintenance costs for each of the major pieces of equipment in the glass separation system were estimated by suppliers of the equipment.

The estimated maintenance costs are about 6 percent of the uninstalled equipment cost per year. The installed equipment cost is given in Eq. 3. By accounting for the adjustment for the installation of 75 percent and by assuming 2000 hours per year of operation, the unit cost of maintenance is calculated to be

$$UCM_G = 0.4 + \frac{2}{\dot{m}} \quad (19)$$

where:

UCM_G = unit cost of maintenance of a glass separation system ($/Mg); and

$\dot{m}$ = rated feedstock flowrate (Mg/hr).

5.4.4 Non-Ferrous Separation

The key piece of equipment of the aluminum recovery system is the eddy current separator. Most commercial-scale MSW processing plants do not include eddy current separators (ECS), and some of the plants that have ESC units have stopped using them. Only one manufacturer provided data which CRS considered reliable and reasonable. For a 6 Mg/hr aluminum separator, the manufacturer specified 2 hr/wk of maintenance labor and 2.5 percent of the capital cost ($120,000) per year for supplies. At a wage rate of $30/hr and 2,000 hours of operation per year, this yields a unit maintenance cost of $0.50/Mg of feedstock to the eddy current separator.

The remainder of the aluminum recovery system comprises a trommel screen preceding the eddy current separator and an air knife following the separator. About 80 percent of the mass entering the system is removed by the screen. Therefore, the unit maintenance cost of the eddy current separator is $0.10/Mg of feedstock to the system. The screen costs about $.20/Mg to maintain. The air knife receives as feedstock about one thirtieth of the feedstock to the system. Assuming that maintenance of the air knife is similar to that of an air classifier ($.30/Mg), its unit cost is about $0.01/Mg of input to the system.

The total unit maintenance cost for the aluminum separation system is taken to be $0.30/Mg.

5.5 RESIDUE DISPOSAL COSTS

The residue disposal costs are modeled on a plant-wide basis rather than on a unit process basis because the output from one unit operation is often the feedstock to another unit operation. The unit cost of residue disposal is,

$$UCR = mf_r \times C_r \qquad (20)$$

where:

UCR = unit cost of residue disposal (\$/Mg of feedstock to the system);

mf_r = mass fraction of MSW feedstock that is disposed; and

C_r = cost of disposing residue (\$/Mg).

The cost of disposal, C_r, includes the cost of transporting the residue from the processing plant to the disposal site as well as the cost of disposing it. The mass fraction to be disposed (mf_r) is an input to the economic model and is determined from the mass balance model. It is equal to one minus the sum of the mass fractions recovered as salable materials.

5.6 REVENUE

The income from the sale of recovered materials is modeled on a plant-wide basis because allocating revenues to a particular unit operation is arbitrary when more than one unit operation contributes to the recovery and upgrading of the material. The unit revenue is

$$UR = (mf_{fe} \times P_{fe}) + (mf_{rdf} \times P_{rdf}) + (mf_{nf} \times P_{nf}) + (mf_g \times P_g) \qquad (21)$$

where:

UR = unit revenue from the sale of recovered materials (\$/Mg of raw MSW feedstock);

mf_{fe} = mass fraction of raw MSW recovered as ferrous scrap;

mf_{rdf} = mass fraction of raw MSW recovered as refuse derived fuel;

mf_{nf} = mass fraction of raw MSW recovered as non-ferrous scrap;

mf_g = mass fraction of raw MSW recovered as glass;

P_{fe} = price of ferrous scrap, FOB the processing plant; and
P_{rdf} = price of refuse derived fuel, FOB the processing plant; and
P_{nf} = price of non-ferrous scrap, FOB the processing plant; and
P_{g} = price of glass, FOB the processing plant.

The mass fractions are determined from the mass balance model presented in this report and in the Task 3 Report. The prices are inputs to the model and are assigned by the user of the model.